PROCESSING IMAGES OF FACES

Tutorial Monographs in Cognitive Science

Nigel Shadbolt, Series Editor

Processing Images of Faces

edited by

Vicki Bruce & Mike Burton

ABLEX PUBLISHING CORPORATION
NORWOOD, NEW JERSEY

Library of Congress Cataloging-in-Publication Data

Processing images of faces / edited by Vicki Bruce & Mike Burton.
    p.   cm. — (tutorial monographs in cognitive science)
    Includes bibliographical references and index.
    ISBN 0-89391-684-6 (Cl). — ISBN 0-89391-771-0 (ppb)
    1. Optical pattern recognition. 2. Image processing. 3. Face perception. I. Bruce, Vicki. II. Burton, Mike, 1959-
III. Series.
TA1650.P76   1991
621.36'7—dc20   90-27403
                                   CIP

Ablex Publishing Corporation
355 Chestnut Street
Norwood, New Jersey 07648

This book is dedicated to the memory of Frank Burton

Contents

Acknowledgments

In the Spring of 1988, Vicki Bruce, in association with Hadyn Ellis and Andy Young, organized a weekend workshop at Grange over Sands to which we invited psychologists, computer scientists, and engineers who were involved with image processing applications using faces. The 1988 workshop gave us the idea for the collection we have here edited. It was clear from the enormously successful meeting at Grange over Sands that there was much common ground between scientists from different disciplines working on face processing, and the current collection reflects this. The collection is not, however, an edited conference proceedings. While many (not all) of the authors attended the Grange over Sands meeting, their contributions here include general reviews of their particular research area, as well as specific discussions of ongoing research. We hope that the book will therefore provide a much-needed tutorial overview of current research on facial image processing.

The weekend which inspired this book was funded by the Economic and Social Research Council as one of an annual series of workshops (1984–1990) which have brought together British researchers to discuss different aspects of face perception. These workshops have helped foster collaborations between different groups of scientists working in face processing, a collaboration which culminated in an ESRC-funded research programme (ref XC15250000) which links research on face recognition at the Universities of Aberdeen (Ian Craw), Cardiff (Hadyn Ellis), Lancaster (Andy Ellis and Andy Young), Nottingham (Vicki Bruce), and St. Andrews (Dave Perrett). The weekend workshops from 1988 have been funded as part of this program, and we would like to thank the ESRC for encouraging and funding our collaborations in this way. Preparation of the manuscript has been supported in part by the ESRC research program, and additionally by research grants from the SERC Image Interpretation Initiative to Vicki Bruce and Mike Burton, one of which now funds a collaborative project with Alf Linney at University College, London. This collaboration also grew out of our meeting at Grange over Sands.

Preparation of this volume has been aided by our excellent computing facilities in the Department of Psychology at Nottingham University. We would like to thank the series editor, Nigel Shadbolt, for his encouragement, and the support staff at Ablex for help in the final stages of preparing the manuscript.

Vicki Bruce and Mike Burton
November 1989

Preface: The Human Face—The Organ of Communication

Vicki Bruce
and
Mike Burton
Department of Psychology
University of Nottingham
Nottingham, England

The human face provides information which regulates a variety of aspects of our social lives in ways of which most of us are scarcely aware. Expressive movements inform us of the emotional state of our companions, and other facial gestures help to regulate and punctuate conversation. Communication, particularly in noise, is also helped by our perception of lip and other movements associated with articulation. Visual speech neatly complements acoustic speech, as phonetic features which are difficult to hear in noise are readily visible (see Brooke, this volume). Additionally, the face serves to identify its owner. While people can be identified by other means such as voice, gait, or clothing, the face is the most reliable key to individual identity. Face identification is a particularly interesting variety of object recognition, because the face must be identified despite the complex nonrigid motions associated with its other social functions.

It may have been this seeming complexity of social function mediated by the face which initially hindered theoretical progress in understanding *how* meaning is derived from the face (e.g., see Ellis, 1975, for an early review of the area). Research during the 1970s was largely inspired by some much-publicized cases of mistaken eyewitness identification, but this research produced little more than a catalogue of empirical findings. The past few years, however, have witnessed remarkable progress by cognitive psychologists in understanding at a broad, "macrostructural" (Rumelhart & McClelland, 1985) level, the functional components which are involved in face perception, and the interrelationships between these functional components (e.g., see Bruce, 1979, 1983; Bruce & Young,

1986; A. Ellis, Young, & Hay, 1987; H. Ellis, 1986; Hay & Young, 1982; Young, Hay, & Ellis, 1985).

A broad functional model of face processing (e.g., see Bruce & Young, 1986) has resulted from converging evidence from experimental studies with normal adults, everyday difficulties in recognizing people, and investigations of the deficits in face processing which may result from brain injury. It has been established that the processing of facial expressions and facial speech proceed in parallel with the identification of faces. Facial identification itself involves discrete stages which proceed in a sequence from perceptual classification (recognizing different views of a familiar face as familiar) through semantic classification (knowing that the familiar face is a politician) to name retrieval (knowing that a familiar face is George Bush). A number of recent volumes summarize theoretical, empirical, and applied research developments in this area (e.g., see Bruce, 1988; Bruyer, 1986; Davies, H. Ellis, & Shepherd, 1981; H. Ellis, Jeeves, Newcombe, & Young, 1986; Young & H. Ellis, 1989), and the current volume does not attempt to review this growing field of cognitive psychology. Rather, this book reviews developments within the growing interdisciplinary "cognitive science" of facial image processing.

Ongoing research by psychologists in this area takes the broad functional model as a framework for guiding research aimed at elucidating the microstructure of the face processing system—at examining how such processes as "perceptual classification" actually take place. Theoretical developments in this area are increasingly dependent upon computer technology, both because image-processing techniques allow us to display and manipulate faces for experiments in ways which were not feasible with old photographic technology (e.g., see Benson et al., Bruce et al., this volume), and because theoretical ideas can be expressed and tested more rigorously using computer simulation (e.g., see Rolls, this volume). The psychological contributions in the current volume have been chosen to illustrate current theoretical developments which are heavily dependent upon image processing and computer simulation.

As technology improves, so it becomes increasingly feasible to automate many aspects of face processing that humans take for granted (e.g., see Craw & Tock, Brooke, this volume), and to develop new technological aids. This book includes examples of computing developments for forensic purposes (Shepherd & Ellis, this volume), for the simulation of plastic surgery (see Linney, this volume), and for automatic animation for applications in telecommunications and creative arts (see Duffy, Waters, this volume). Knowledge of human face perception can provide important constraints and insights in the development of these new technological aids, as the chapters in the second part of this book demonstrate.

For these reasons, then, the study of face image processing is an increasingly interdisciplinary topic. In this book we have drawn together contributions by psychologists, computer scientists, and engineers. Each group makes a different but complementary contribution to our understanding of how the brain processes and recognizes faces, and/or to the development of technology for applied problems of face reconstruction and recognition. Here we provide a short overview of the contents of the two parts of the book.

PART I: FEATURE EXTRACTION AND FACE RECOGNITION

All faces must be similar in overall configuration to subserve their various functions. A face bears sensory organs in appropriate locations and allows us to breathe and eat, and these basic functions constrain the form of the face, so that individuating differences must be subtle. The human brain finds it relatively easy to recognize different identities from these subtle differences, yet automatic face recognition poses a considerable challenge to computer scientists and engineers.

There have been two rather different approaches to automating face recognition (see Bruce & Burton, 1989, for a more extensive recent review). Abstractive approaches seek to define a set of key parameters for the measurement and matching of faces. To automatize such measurements involves finding ways to locate facial features within images—a problem which has proved remarkably difficult. Craw and Tock describe some of these difficulties in their review of work aimed at locating and measuring face features in a facial image. Brooke outlines a similar problem in reviewing work aimed at measuring the shapes of mouth movements for the synthesis of facial speech.

In contrast to these abstractive approaches, some recent network implementations (e.g., see Stonham, 1986) solve the face recognition problem by brute force (in the form of massive memory) and ignorance of any knowledge of facial features or dimensions. While such simple pattern associators may be successful in constrained situations, as Craw and Tock point out in their chapter, they cannot readily be applied in situations where such knowledge about faces must be made explicit. Such criticisms do not necessarily apply to more complex, multilayered connectionist architectures, where features may be discovered by hidden units or competitive learning algorithms, and so on (see reviews in the collections edited by Rumelhart & McClelland, 1986). Rolls describes how the responses of single cells in the monkey's cortex to faces can be interpreted as arising from the ensemble encoding of facial identity, in which identity is encoded as a pattern of activity distributed across a relatively small ensemble of processing units. He also considers how the observed tuning of these neurones might arise through a competitive learning mechanism.

Theories of the internal representation of faces by humans should be able to explain the apparent paradox of caricature—where a metrically distorted version of a face seems to provide a better key to identity than the veridical face itself. While previous work on caricatures has used line drawings, Benson, Perrett and

Davis describe the development of a system to produce automatically caricatured photographs. This work illustrates very well how sophisticated computer graphics can allow scientists to ask questions about human perception which were impossible with old technology. Finally in this part of the book, Watt provides an overview of the kind of information needed to categorize the face in different ways (as a face, as a female face, as a smiling face, and so forth), and discusses the way in which early visual processing of faces by the human visual system may deliver descriptions nicely appropriate to the needs of an observer at different distances away from another person.

PART II: FACIAL RECONSTRUCTION AND ANIMATION

A number of quite distinct applied problems require that faces be reconstructed. For investigative purposes, the police may require a witness to try to build an image of a wanted person. Shepherd and Ellis describe the development and evaluation of interactive electronic systems for building a face from a set of remembered facial features. Their chapter also reviews the development of a system for automatic retrieval of mugshots from a witnesses' description. Such an automatic system would reduce the number of mugshots witnesses need search through, and hence increase the chance that they might recognize a reoffending criminal without interference from the sight of hundreds of other faces. Interestingly, for such a system to work in practice, some means of automatically measuring existing mugshot files must be devised—the very problem which motivated Craw and Tock's project described earlier in the book.

Advances in facial surgery mean that it is now possible to remodel the face of a deformed or severely injured person. Linney describes the development of a system to allow such surgical interventions to be simulated electronically on X-ray images of the patient's bones. A representation of the face is obtained by superimposing an image of the skin layer obtained pre- or postoperatively from a laser scanner. The development of such techniques for measuring the shape of the face provides important data for other projects, as illustrated by subsequent chapters in this part of the book.

Linney models the face as a bumpy surface rather than a two-dimensional pattern. The use of 3D face representations becomes essential in the automation of certain animation techniques. Duffy addresses the problem of reconstructing and animating an image of a moving, expressive face for the purpose of developing a video telephone, while Waters describes how the animation of realistic facial expressions has been made possible by considering the anatomy, mechanics, and control of facial musculature. Finally, Bruce, Burton, and Doyle describe how the use of three-dimensional models of the face and head is helping in the development of a theoretical approach to face perception which is based upon a proper understanding of the structures to be viewed. Faces are not flat patterns, but three-dimensional shapes produced and constrained by growth.

REFERENCES

Bruce, V. (1979). Searching for politicians: An information-processing approach to face recognition. *Quarterly Journal of Experimental Psychology, 31*, 373–395.

Bruce, V. (1983). Recognizing faces. *Philosophical Transactions of the Royal Society of London, B302*, 423–436.

Bruce, V. (1988). *Recognising faces.* London: Erlbaum.

Bruce, V., & Burton, M. (1989). Computer recognition of faces. In A.W. Young & H.D. Ellis (Eds.), *Handbook of research in face processing.* Amsterdam, Netherlands: North-Holland.

Bruce, V., & Young, A. (1986). Understanding face recognition. *British Journal of Psychology, 77*, 305–327.

Bruyer, R. (Ed.). (1986). *The neuropsychology of face perception and facial expression.* Hillsdale, NJ: Erlbaum

Davies, G., Ellis, H., & Shepherd, J. (1981). *Perceiving and remembering faces.* London: Academic Press.

Ellis, A.W., Young, A.W., & Hay, D.C. (1987). Modelling the recognition of faces and words. In P.E. Morris (Ed.), *Modelling cognition.* Chichester, England: Wiley.

Ellis, H.D. (1975). Recognising faces. *British Journal of Psychology, 66*, 409–426.

Ellis, H.D. (1986). Processes underlying face recognition. In R. Bruyer (Ed.), *The neuropsychology of face perception and facial expression.* Hillsdale, NJ: Erlbaum.

Ellis, H.D., Jeeves, M.A., Newcombe, F., & Young, A. (1986). *Aspects of face processing.* Dordrecht, Netherlands: Martinus Nijhoff.

Hay, D.C., & Young, A.W. (1982). The human face. In A.W. Ellis (Ed.), *Normality and pathology in cognitive functions.* London: Academic Press.

Rumelhart, D.E., & McClelland, J.L. (1985). Levels indeed! A reply to Broadbent. *Journal of Experimental Psychology: General, 114*, 193–197.

Rumelhart, D.E., & McClelland, J.L. (1986). *Parallel distributed processing: Explorations in the microstructure of cognition.* Cambridge, MA: MIT Press.

Stonham, J. (1986). Practical face recognition and verification with WISARD. In H.D. Ellis, M.A. Jeeves, F. Newcombe, & A. Young (Eds.), *Aspects of face processing.* Dordrecht, Netherlands: Martinus Jijhoff.

Young, A.W., & Ellis, H.D. (1989). *Handbook of research in face processing.* Amsterdam, Netherlands: North-Holland.

Young, A.W., Hay, D.C., & Ellis, A.W. (1985). The faces that launched a thousand slips: Everyday errors and difficulties in recognising people. *British Journal of Psychology, 76*, 495–523.

I
Feature Extraction and Face Recognition

1
The Computer Understanding of Faces*

Ian Craw
David Tock
Department of Mathematical Sciences
University of Aberdeen
Aberdeen, Scotland

INTRODUCTION

In this chapter we describe a computer program called "GetFace," which attempts to see and understand a human face. The input for our program is obtained by pointing a black-and-white video camera at a face. The camera splits the resulting image into a number, typically 512×512 or $262,144$, of squares or pixels from each of which the averaged grey level is recorded. This list of numbers is the data on which our program operates. The aim is to recognize and measure features, such as the outline of the head, the eyes, nose, and mouth.

Our program is knowledge based. In addition to the pixel information about the current face, it has information about the average face and explicit knowledge about what to expect within a face. This takes the form of statements like "the mouth lies below the eyes," containing knowledge of the relationship between the individual components of the face. We describe work in progress. The design and underlying structure have been fixed, but as yet only some of the components have been written and tested. We are also working at present only with full-face images, and have ignored complications such as beards and glasses.

Our motivation and viewpoint is that of an engineer; we wish to construct a working system with the currently available technology. We present in this chap-

---

* This work was supported in part by SERC research grant number GR/E84617 to Ian Craw (Department of Mathematical Sciences, University of Aberdeen) and Roly Lishman (Department of Computing Science, University of Aberdeen). We are grateful to Dr. Denis Rutovitz, of the MRC Clinical and Population Cytogenetics Unit, Edinburgh, for supplying a copy of the SBS system discussed here. Finally, we thank Roly Lishman for many valuable conversations in which some of the ideas discussed here were first considered.

ter detailed results of our design process. We hope this will be of interest to others working within computer vision. In addition we hope to attract the attention of those concerned with modeling the human face perception system.

One reason to model the human system is as an aid to theory building. Higher-level theories of face perception are designed to explore the various logically possible approaches consistent with known experimental results. Different choices within a model often lead to different intermediate results. As such it is often possible to devise experiments to explore the point. For example, in the model published by Bruce and Young (1986), modules concerned with expression analysis and deciding whether a face is familiar operate independently, but each of these functions occurs after processing by a structural encoding module. In this example a number of experiments have been done to justify these orderings. As the main features become clear, individual components can themselves be investigated, and so the top-down modeling proceeds.

The program GetFace itself is not meant to be a model of the human face recognition system in the above sense. However, compared with these models, GetFace is described in very considerable detail. We hope our description may serve to widen the modeling vocabulary, and that some of the lessons learned from machine perception may transfer. For example, we have little formal ordering of tasks in our system; those orderings that are used are the result of experience and are chosen on the basis of efficiency. It is also possible that individual components may have similarities.

People can routinely recognize faces, and it seems likely that computers will soon be able to do so. An analogy might be drawn with the relationship between the flight of birds and artificial flight; each perspective illuminates the other, although different available technologies often enforce different choices of method.

We give an example of feedback from the computer to the human domain. At a recent meeting at Grange over Sands, some studies were reported informally of a "split brain" patient who had gross communication difficulties within the brain. When tested with tasks involving recognizing parts of a face in unusual locations, the errors made were not immediately explicable. For example, images of a face in which the positions of the nose and mouth had been interchanged were viewed in turn by the left and right fields. In one case the features were misidentified as being in their usual positions, while with the other field, the interchange was detected. It seems as though the outline of the face was being correctly processed, but that the internal features were assumed to be in their default positions. For comparison, we can ask whether GetFace could have the same difficulties. In fact, were a breakdown in communication to occur at a similarly high level within the GetFace program, say between the "outline" and "internal features" modules, similar misidentifications could well be made. This similarity suggests that such errors could occur with a wide variety of possible models and so should not be regarded as unusual.

Our interest in this topic arose from a collaboration between Aberdeen psychologists and the Home Office on FRAME (Face Retrieval And Matching Equipment), essentially a computerized mugshot system for enabling witnesses to identify a face of someone with a police record (see Shepherd & Ellis, this volume). An experimental form of this comprises a database of 1,000 male faces, each coded on a large number of feature size and interfeature distance measures, together with information about coloring. The witnesses' description of a suspect's face is first coded in a suitable fashion. The computer then searches the database of faces, ordering them according to their similarity to the description. Typically, the first half dozen faces are then shown to the witness; if none of these is the culprit, he or she may suggest mixing features from different faces, and the computer can then search on the new set of measures. This process of refinement can continue until the witness spots the suspect's face; at present overall success rates of more than 80% are being achieved, suggesting that the current size, distance, and color measurements are usually enough to separate faces.

As part of the system, detailed measurements of individual facial features are required. Because the number of faces involved was potentially very large, we looked at the possibility of taking these measurements automatically. This is reported in Craw, Ellis, and Lishman (1987); our present work grew out of this study.

In some sense, the task is easy; people can find the mouth within a face rapidly even on badly degraded images. Computers are much less successful; at present we attribute this failure to an inability to integrate knowledge from different sources. So, for example, we can write an algorithm which is good at recognizing a mouth, provided it is only searching in an area which is almost correct. We find it much harder to incorporate our knowledge of where to look for individual features. This knowledge comes from our experience of recognizing many faces, but is not easy to make explicit, except in terms of relationships with other features.

Recognizing objects from one or more digitized views falls within the domain of computer vision, and below our work is placed within this context. At this stage we note there is physiological evidence (see Rolls, this volume) consistent with the view that face processing in humans takes place in special purpose modules; certainly it is a specialized and frequently performed task. This is seen as a justification for developing special purpose face processing modules, rather than hoping for acceptable results from general computer vision programs. Indeed, such a specialized module would be of use in more ambitious vision systems: working in parallel, it may simply be necessary to actuate the face module whenever a suitably rounded shape is identified.

Practical Applications

So far we have indicated why GetFace may be of interest to those modeling human face recognition, and have outlined the computational background. It seems likely that the system will have applications at a fairly early stage in its development, even though it would then be neither particularly accurate nor very fast.

We noted a potential use by the police for back-record conversion on existing mugshots. Field trials of the underlying FRAME system (now known, perhaps more judiciously, as FACES) are taking place at present, using data that were measured by hand from the photographs. In a larger system it would be necessary to take the measurements automatically. However, absolute accuracy is not essential, as long as the system knows when failure has occurred; the missing measurements can, if necessary, be taken by hand. There is also no very pressing need for speed, since other transactions, entering the name and address of the person concerned and making other judgmental ratings, must take place in parallel.

Other likely applications include credit card verification and security locks. People expect to be recognized by their faces: no great change in habits would be required if a credit card carried a compressed representation of the holder's face, and a check was made against the face of the carrier when the card was used. Again a processing time of 20 seconds is likely to be acceptable, since other operations associated with the sale must take place in the same time.

A face recognition scheme would also make a convenient system of access control in secure buildings. A person would be identified on entry to the building, and an encoded description of his or her face stored. Thereafter, doors would open for him or her only if his or her face matched this stored description and he or she had been authorized to enter that area. More demanding would be other forms of access control, for example to football grounds. Here again the check would be that the membership card contains compressed details of the face of the person presenting the card at the security barrier. However, a judgment would have to be made within a very few seconds, at which point those not recognized would be segregated for further screening. This puts a much greater premium on speed and accuracy.

Uses where matching is undertaken with a stored copy of the user's face assume the user is willing to cooperate with the system. It may be necessary to present a different, perhaps more relaxed, view of his or her face should the system fail to recognize it initially. This need for cooperation is present in many currently useful speech recognition systems where it is not too great a handicap.

Another form of access control is the exclusion of specific people whose faces are known. In this case, the person would be trying to defeat the system by not being recognized. It seems certain that this would always be possible, given a sufficiently elaborate disguise, but some clues would be available from the system.

PREVIOUS WORK

Computer vision started with the image enhancement of early satellite photographs and quickly progressed to simple object recognition. Driven by applications, much effort has been expended on certain sorts of images. One of these, important for automated assembly, consists of machine parts. The edges are usually simple geometrical curves; these may be of many shapes, but parts of the same shape are essentially identical. With control over the lighting, simple methods have often proved successful here. A much harder vision problem, which has had significant funding, is that of target recognition, particularly in cluttered outdoor scenes. Within the biological or medical domain, computer vision has slightly different aims; usually the significance lies in the small differences between different instances of the same class of object. Again, fairly simple systems are successful, and many systems are now in constant use doing routine cell analysis.

Vision techniques can be simple; often a suitably chosen threshold permits recognition using decision tree methods or template matching. However, as the task becomes more complicated, it seems necessary to use additional world knowledge. This is usually done for two reasons: to improve overall control by incorporating feedback as appropriate, and to reduce potentially large search spaces when matching. Knowledge is often incorporated in the form of models; the aim is, first, to divide the image into components that have been modelled, and then to use constraints imposed by these models and the way they combine to restrict feasible interpretations in the image.

To begin the necessary matching process, local information has to be obtained directly. Much of this is standard; thresholding, edge detection, region growing, and texture analysis. It can include local shape information, perhaps obtained as "shape from shading," or in other ways, and leading to the "primal sketch" or local surface description of Marr (1982). Other intermediate results have been proposed, such as the "curvature primal sketch" of Asada and Brady (1986). Even at this level knowledge has been used; Nazif and Levine (1984) describe an expert system for low-level image segmentation based on a number of general rules. They show it significantly outperforms more conventional methods, albeit in a run time of 30 minutes.

These techniques have been combined into working systems. Binford (1982) surveys 10 knowledge-based imaging systems, including those for recognizing aircraft on the ground, objects on a desk, and analyzing urban scenes. Within the biological domain, Niemann et al. (1985) describe a diagnostic system applied to images of the human heart; here the vision ideas are fairly simple and specialized, but again the system is effective. A number of these systems have individual components that are very sophisticated; nevertheless, it seems that none has proved useful outside its initial domain. Certainly, no general computer vision system is even on the horizon at present.

None of these systems seems able to "understand" faces or similar objects, where the aim is to distinguish between different faces. However, there has been some specialized work in this area. An early successful approach to recognizing facial features was that by Sakai and colleagues (Sakai, Nagao, & Fujibayashi, 1969; Sakai, Nagao, & Kanade, 1972); further progress is reported by Kanade (1977). Their method finds the features in a specified order; the sides of the face before the nose, mouth, and chin, which in turn are found before the eyes are sought. However this serial control structure has feedback; failure at a given stage can lead to a reexamination of the previous results. Most of their identification is done on the image obtained by first applying an edge detector and then suitably thresholding the output to get a binary image. Pixels in a narrow window were then averaged into a single representative row or column, and this was used to detect features such as the edge of the face. These methods successfully located the head top, face sides, nose, mouth, and chin contour, achieving success in 607 out of a test set of 800 faces.

In an attempt to reduce the search area involved for large images, Kelly (1971) used a "planning" or multiresolution technique. Here the required object is located at small scale, and the information used to guide a full scale search. He illustrated the technique by searching for face outlines, demonstrating significant reductions in search times. A number of workers, such as Baron (1981), have concentrated on face recognition without having individual feature recognition as a subgoal.

Little recent work on faces has appeared, even though hardware advances allow the use of techniques which would have been prohibitively expensive to earlier workers. A preliminary study by Craw et al. (1987) confirmed the value of multiresolution methods but was still governed by hardware limitations. These prevented the feedback of information from finer to coarser resolutions in order to improve the initial locations of limited search areas.

An alternative approach to recognition is by using a system that can learn, training it with examples of the object to be recognized. This technique is adopted, for example, by WISARD (Stonham, 1986); it can work equally well on both faces and features within a face. With the current popularity of neural net algorithms, a number of other systems are being built with these aims; learning appropriate positional relationships between eyes, nose, and mouth features is being used by British Telecom as one of several test problems used in evaluating neural net systems. A disadvantage with this approach occurs when measurement is attempted. A mouth might be recognized, because it is sufficiently similar to example mouths on which the system has been trained. However, the extremities of the mouth are not themselves understood well enough for the length of the mouth to be measured. Thus, if measurements are needed as output, the system has to be trained to make these measurement initially. With more principled methods, measurement can easily follow identification.

ARCHITECTURES FOR VISION

All the work we described above on recognizing facial features had control mechanisms which were at most a minor modification of a strictly serial control structure in which, for example, the eyes are found first, then the nose, and so on in a predetermined order. Such a system has a number of problems associated with it, the main one being the liability to a catastrophic failure. This occurs when a successful identification is rejected at too early a stage in processing, before enough evidence for its correctness has accumulated. Since we are trying to identify a few features, and in general there are many candidates for each feature, such a premature rejection is likely to lead to a complete failure of identification. An additional problem arising from the interdependence of the various parts of the program is the difficulty of changing both the individual algorithms and the control strategy. Indeed this flows from the effort needed to hand craft the program to suit the particular application.

Ideally we would have a number of methods available. Each method should be capable of being applied in an opportunistic manner, and each should give a clear indication of success or failure. We thus seek a more flexible architecture in which knowledge about individual features is separate from the underlying control strategy. We expect our experience in writing recognition algorithms to grow as we develop the program. A transparent representation of the underlying knowledge is thus desirable, so that it may easily be changed as new strategies are adopted. Because of this, we choose at this stage to rule out example-based recognizers such as WISARD (Stonham, 1986), even though, at a later stage, we may choose to incorporate them as independent knowledge sources. Finally we seek a very flexible control strategy that can use all the information given by a piece of world knowledge. Thus the information that "the mouth lies below the eyes" should, in different circumstances help with locating either the eyes or the mouth; indeed, if both of these have already been located, it should provide a check on the meaning of "below."

An additional difficulty that can occur in computer vision systems is an unreasonably slow execution speed. Although this may not be relevant during development, the design should be capable of running much more rapidly. We interpret this as a need for making explicit the interdependencies between the various parts of the system, so a parallel implementation becomes possible.

These problems of flexibility, modularity, control, and concurrency have been recognized with increasing frequency. A popular solution is that of the black-board system. Such a system, Hearsay II (Erman, Hayes-Roth, Lessor, & Reddy, 1980), was implemented for certain speech-recognition tasks. The architecture rapidly became popular, and was soon used in computer vision (Draper, Collins, Brolio, Hanson, & Riseman, 1988; Nagao, Matsuyama, & Mori, 1979). A recent survey of more than 30 such systems can be found in Engelmore and Morgan (1988).

A blackboard system is a way of organizing a collection of experts, or knowledge sources, with different competencies. It is based on a model of problem solving in which the experts gather round a blackboard and cooperate in an opportunistic manner to work towards a solution. Initially, the blackboard contains only a statement of the problem or initial goal. As the experts read the data on the blackboard, they contribute to the global understanding of the problem by writing additional information on the blackboard for all to see and process. This additional information may be in the form of starting conditions, additional relevant world knowledge, or specialized processing of the input data. Alternatively, it can be in the form of subgoals which one of the experts considers are worthwhile trying to achieve. As the solution progresses, the content of the blackboard is modified; subgoals are achieved, hypotheses verified or rejected, and ambiguities in the initial specification investigated.

This organizational model has a number of attractive features. One is modularity; additional experts can be added without affecting those already present. Another is the support for both procedural and declarative knowledge, and both top-down (or goal-directed) and bottom-up (or data-driven) reasoning can be used. Also, since each expert is autonomous and all have access to the blackboard, there is no bar to a parallel implementation in which experts that can contribute at a given stage are fired simultaneously.

A Simple Blackboard System

A simple blackboard system, SBS, has recently been described by Baldock and Towers (1988). It is written in POP11, and the control structure at its core consists of a simple goal list, treated as a stack. Each knowledge source, or expert, has a collection of goals which it can in principle satisfy. In attempting to do so, there may be preconditions which must be met, and these preconditions, when executed, may themselves place further goals on the goal list, hence, implementing a form of backward chaining. When the preconditions have been met, the actions to be taken by the expert are governed by simple production rules. These rules may, however, call arbitrary POP11 procedures. The effect is that the knowledge within the expert is easily visible, while at the same time data-driven inference is permitted.

The POP11 database is used by the system as part of the blackboard. This is a simple list which is initially empty. In order to record information in that list in a uniform manner, data is written to it in a frame-like data structure (Towers, 1987) which, in principle, consists of labeled collections of slots or item-value pairs. We describe one such frame, used to store details about a point, in the next section. In fact the implementation is significantly richer than this; the values are not necessarily calculated until they are needed, and it is possible to have default values for a slot, while a daemon may if necessary sit in a slot, to be invoked whenever a value is requested. Finally, an effective notion of inheritance is supported, allowing new frames to be constructed easily. The blackboard then consists of this database, together with global POP11 variables, and the goal list described above.

Control is in principle simple. If the goal list is empty, the system exits indicating success. Otherwise an agenda is collected containing those experts which can satisfy the goal at the head of the goal list. This agenda is then sorted by the priorities currently assigned to the experts. If the agenda is empty, the system indicates failure; otherwise, the highest priority expert is executed and a new agenda gathered. The power of this control structure lies in the ability of both the rules and the preconditions within an expert to modify the goal list itself, and the priorities of individual experts.

One of the attractions of such a system is its simplicity; it is implemented in POP11 in a few hundred lines of code, mainly as a collection of macros for easing the use of the "sections" facility. However, there are problems with this simplicity; only one expert will be consulted to satisfy a given goal, even though several claim to be able to do so, and it must complete successfully or the system reports failure. It is thus harder for a number of different experts, with similar competencies but using different methods, to cooperate in generating a solution. Another problem arises because the control is purely goal driven; even though the state of the blackboard has been changed, the current goal on the blackboard drives the choice of the next expert to be selected, rather than a more "obvious" but opportunistic choice.

THE PROGRAM

The program GetFace is built using the structures of SBS just described. The aim is to process the digitized image of a face, and produce locations and measurements of features within the face, and it uses a collection of cooperating experts which communicate by writing data to a blackboard. However, we modify the control mechanism by placing the control itself in the hands of another expert. While it is not necessary to modify SBS in this way, it seems more flexible to do so than to insist on a "control GetFace" goal as the initial head of the goal list.

Although SBS provides us with an overall structure, it makes no direct contribution to the fundamental problem of computer vision, that of modeling the objects to be recognized in a form suitable for matching with observation. In GetFace we build these models within the feature experts described below. In order to do this, we need a language which is in part built from the primitives of lower-level experts and is in part concerned with relationships.

Our philosophy is that each individual feature can be accurately located using standard vision tools. The main difficulty is an impractically large search space when the likely outputs from all our feature detectors are combined. We reduce this search space drastically by using the known interrelationship between the features. We shall thus be seeking a feature in the "correct" place; as the program progresses, partial identifications will both eliminate many subsequent potential feature locations, and also confirm (or otherwise) the identifications previously made.

In order to express relationships in a suitably fuzzy way, we use a multiresolution approach to geometry—see, for example, Klinger (1984)—and consider objects as existing in principle at all scales, from the coarsest, in which the whole image is represented as a single pixel, to the finest full-scale description. We use this as a framework, enabling a feature to be described and located at a suitably coarse scale before its location is confirmed in detail.

To fix ideas, we consider the case of an image with 512 rows of 512 columns, with both rows and columns labelled from 0 to 511. The two points (150,150) and (255,255) are well separated at this scale. However, when the resolution of the image is reduced by combining squares of four pixels into a single pixel, their separation is reduced. Indeed they both coincide with the pixel (1,1) when the image is coarsened to have only 4 rows of 4 columns, since each original coordinate lies between 128 and 255. This example does not generalize; the point (255,255) does not coincide with (256,256) until the image is coarsened to be just a single pixel. We are seeing the effect of a particular granularity, rather than a more useful generalized granularity. The remedy involves extending our notion of equality and regarding two points as identical at a given scale if they are either the same, or are adjacent pixels. With this extension, nearness becomes equality at a suitably coarser scale.

To illustrate the utility of this approach, consider the interpretation of a description such as "the right eye lies above the right corner of the mouth." Even with an accurate notion of *above,* this is unlikely to be *exactly* true. However it is approximately true, and if interpreted as meaning that both features lie on the same vertical line, becomes exactly true at some coarser resolution.

We illustrate how we express these ideas within a frame data structure. We have the declarations:

```
frame quad isa object;
    slots name, type, resolution, coarser, finer = [];
endframe;
frame point isa quad;
    slots x, y, linelist = [];
endframe;
```

Here an object is a universal data type, containing a unique creation time and other accounting information. Built from this is an abstract piece of resolution-linked structure we call a *quad*. It is designed to link parts of the image at various scales, and is based loosely on the notion of a quadtree. A quad exists at a particular resolution but is linked to other quads of the same type at adjacent resolutions. It stores the name of the object to which the given object coarsens, and a list of known finer objects. This list is initialized to be empty, and names are added as they become known. A particular example of such a quad, again constructed using the inheritance mechanism of SBS, is the point frame. This has the expected slots for its coordinates, but also stores information about lines at this resolution which pass through it. We can thus rapidly verify whether there is a known line which passes through, or even near, two given points. We have similar line and polyline frames. The description that a polyline is roughly vertical if it coarsens in a suitable scale to (within one pixel of) a vertical line thus becomes both natural to express and easy to verify.

Although a multiresolution approach makes additional bookkeeping demands, we argue that this is justified in additional expressiveness. In fact the overhead need not be great, since calculations at different scales are capable of parallel implementation. There are also analogies with our own visual system, in which an indistinct image with a large field of view is combined with a focus of attention mechanism which provides additional detail in the area examined.

This is the framework within which we place a number of experts which form the body of GetFace. They fall into four classes. Two of these, the low-level *vision* experts, perform standard image processing functions. The actual algorithms for feature detection are contained within *feature* experts, and our aim is that this is the only way in which the overall program is specifically written for faces. Our knowledge about individual features is expressed in the rules which make up the description of each feature expert. Finally we have a miscellaneous collection of *control* experts. In addition to providing overall control of the program within the SBS framework, they implement the language in which task-specific experts negotiate with task-independent experts and are concerned with geometry, housekeeping and recording functions. We now describe the experts in more detail.

Low-Level Vision Experts

We have two categories of low-level vision experts to convert pixel-based descriptions into higher-level (but still image-based, viewer-centered) descriptions.

At the lowest level are the *data* experts, reading only raw data from the original image and writing often voluminous output to the blackboard. Data experts implement traditional low-level image-processing routines. An example of such is detecting the direction and significance of edges within the image by examining the rate of change of the grey levels when moving across or down the image, giving, respectively, a vertical or horizontal edge. This information can be obtained using traditional edge detectors such as the Sobel operator (Pratt, 1978, p. 487), or one of the more modern, but computationally more expensive algorithms such as the Canny edge detector (Canny, 1986). Other information obtainable by data experts includes texture measures and other local statistics, and information in the Fourier domain, where the presence of high frequency energy gives a measure of local "business." In principle these low-level image-processing routines are designed to work rapidly using only information available at or near each point of the image, and can be implemented in hardware, or in parallel if additional speed is needed. In effect, these are versions of the image seen through different filters and are the data that other experts manipulate.

At a conceptually higher level we have *shape* experts. They manipulate the output of the data experts, reading generalized data and writing higher level feature descriptions, together with confidence indicators, to the blackboard. They have no direct access to the underlying data, simply consulting data experts when details are missing. An example of this higher level of abstraction is that of edge linking; data experts report edge pixels, but shape experts link these together, using continuity assumptions to associate apparently related edge pixels, gap spanning where there is good evidence to do so, and reporting the whole as a line. Other shape experts return a list of long lines, perhaps restricted to pass near two or more specified points. Additional constraints can be imposed by investigating only those line segments which agree to a given tolerance with a specified shape, for example, pairs of lines in any long thin horizontal minimum bounding rectangle. In fact, this last shape output is very close to our definition of a mouth at certain resolutions, although by relaxing the requirement on the rectangle somewhat, this shape is also found at many eyebrows.

We have argued that there is a logical distinction between data and shape experts. We also found this distinction useful practically, and so it is mirrored in our architecture. Data experts typically deal with a lot of data, so all requests to a data expert are passed for service to a faster language, normally C. In contrast, our other experts are written in POP11, the same language used to implement the blackboard system itself. To minimize the volume of data passed between the two languages we try to ensure that data to be read by higher level experts for a request concerning an image of size n x n is of order n, rather than of order n^2.

Feature Experts

Both data and shape experts are primarily data-driven; although constructed with our application in mind, they are in principle available to perform any vision task. In contrast, feature experts are the experts which embody the task-specific knowledge describing the individual features sought and their interrelationship. Feature experts are model rather than data driven and are invoked both in response to overall goals and to results placed on the blackboard by shape experts. All vision systems to date are more or less specialized; our aim in locating task knowledge solely within feature experts is to make this specialization as visible as possible. We hope at a later date to give description of feature experts in a language close to "English," relying on other parts of GetFace to translate this into a more amenable form.

Feature experts are organized on a hierarchical basis in order to simplify the control problem. They form a network of nodes in which each node represents a feature at some degree of detail, while the arcs linking nodes represent relationships. Our network is a tree with some additional arcs inserted. The "regular" arcs are those representing inclusion; a feature expert understands the relationship between the subfeatures in its domain. In addition there are relationship arcs, all of which are constrained to be between nodes at the same level, expressing geometrical relationships between features at a given level of detail. These relationships have important implications. For example, we may specify that the eyes lie roughly on a horizontal line, are above the nose, and that the mouth is also horizontal and below the nose. The implication for the relationship between the eyes and the mouth is an important piece of world knowledge and should be available to both the eye and the mouth experts, while each in turn should contribute to updating the notion of "horizontal."

When invoked, a feature expert returns an appropriate list of candidate features. Each candidate has associated with it a collection of evidence for its accuracy, including the location rules that are satisfied and those that are violated. These are all written to the blackboard, and so are made accessible to other experts. However, each feature expert returns a best candidate, based on the evidence available to it when invoked.

We now describe in more detail some of the higher level feature experts in GetFace. In doing so, we shall make extensive use of geometrical notions: at a given stage in the program, the current interpretation of each of these is available from the geometric expert to be described below. At the highest level, a face consists of an oval-shaped outline containing internal features. The oval is vertical, with a typical aspect ratio of 4:3. The internal features always lie inside the outline.

Passing to a finer level of description, the mouth expert is aware that the mouth lies in a long, thin, horizontal box. The mouth is positioned below the nose. The vertical center line of the mouth lies close to the vertical center line of the face. The mouth consists of two almost parallel lips. A vertical line meeting the left corner of the mouth typically passes close to the center of the left eye. A vertical line meeting the right corner of the mouth typically passes close to the center of the right eye. At low resolution the mouth is a single, dark, horizontal bar lying within the mouth-box.

At a finer level still, descriptions of the individual lips, and of teeth, are given with respect to the coarser mouth, although their accurate location will serve to confirm the location of the mouth itself.

The Control Expert

In addition to this hierarchy of vision experts, we mentioned the need for other experts. The control expert arbitrates between competing experts. It has to select at each stage the next expert to be consulted, and so is responsible for an appropriate ordering of the goal list. In order to achieve this, the basic SBS control mechanism is modified to return control to this expert whenever another expert terminates.

The control expert must also be able to resolve problems of conflict and convergence: detecting when convergence is occurring, terminating the program at a suitable point; and when the program is looping, perhaps alternating between two hypotheses neither of which has adequate support, terminating the program and indicating failure. In order to facilitate this, each datum placed on the blackboard is accompanied by a timestamp, specified in our "object" frame above. A nearly static situation can be detected by a feature expert returning an indication of the amount of change. Detecting looping, and yet provoking the reexamination of evidence for decisions which have previously been accepted, is significantly harder. This is an area we have yet to explore in detail.

Speed of execution is not a prime consideration initially. Nevertheless a proper ordering of experts, gathered from the history of previous runs, will lead to better performance. One way to optimize this is to have a "random order" mode in which previous decisions about ordering are ignored. The success or otherwise of these new orderings is then added to the history, and may lead to different decisions during subsequent runs in normal mode. If the random order incorporates natural selection between a number of competing strategies, this is a form of genetic algorithm for the control expert; we envisage this occurring at otherwise idle times in order to build up suitable experience.

Although we downgrade the goal list as a control mechanism, it is still useful in allowing communication between other experts about priorities. Another use is during the development of GetFace to assist testing. During this phase, much more limited goals; "initialise display," "find mouth," and "draw mouth" have value in testing modules within the system.

The Geometric Expert

It is necessary to normalize the images with which we work, which may originally be presented in unusual orientations and at a different scales. For this reason GetFace works in a model coordinate system. The geometric expert maintains on the blackboard the current "best" transformation between image and model coordinates. We refer to this as the *current location transformation*. In particular, there is a current notion of "horizontal" which is used by feature (or shape) experts to limit their search space; changing this will affect their results.

The geometric expert derives the current location transformation by assessing evidence about mismatches between the current best position and the default positions provided by the generic expert. For example if an image is not exactly full-face, the center of the face, taken as the middle of the bridge of the nose, will not coincide with a line drawn equidistant from each side of the face, as fixed from the outline. The difference between these two versions of the center line give a measure of the direction in which the subject was facing when the image was taken. A simple barrel rotation of the image will then be incorporated in the current location transformation.

The geometric expert is also concerned with the positional relationships between objects, and in order to describe these, both in the model and in the images, we need some standard way of referring to subcomponents. For simplicity we give an object a unique position, notionally that of the center of mass of the minimum bounding rectangle, with the assumption that the sides of the rectangle are aligned along the "natural" edges of the image, and an orientation, the direction of the major axis of the object.

The Generic Expert

The generic expert is the repository of our previous experience. It has knowledge of the "normal" or average positions of features within a face, and also of how easy features are to locate. This is useful in two distinct ways: it provides default locations for feature experts; and it can provide "most probable" search strategies for the control expert.

The successful completion of a recognition task will accumulate more knowledge for the generic expert. With many identifications, the stored positional information will become extensive, and we expect to be able to use rather sophisticated statistical techniques. Thus, for example, identifying the outline of the face as "egg shaped" may affect the expected location of the internal features; data on this will be available to the generic expert. The generic expert also stores the information about suitable orderings of the other experts, including those runs in "random order" mode.

Other Experts

A service similar to that provided by the geometric expert is needed in order to assess the evidence for competing hypotheses, and this is provided by the assessment expert. Although decisions are made in the light of the available evidence at a given time, reassessment may become necessary. For example, as the scale and orientation of an image become apparent, potential features which were not accepted must be reconsidered; perhaps a mouth may now be regarded as horizontal. It is thus necessary for the feature to maintain alternative hypotheses.

In order to maintain the modularity of the code, even output is controlled by an independent expert. Although the ultimate goal of GetFace is to produce feature measurements, an obvious way of verifying correct behavior is by displaying a representation of the face as a line drawing superimposed on the original data. The output expert has control of this display device; it can thus fulfill goals during system development such as drawing sets of candidate features at a given level of confidence. More generally it has the responsibility of interpreting the data on the blackboard to the developer of the system.

Testing

We have noted in a number of places in our discussion the interrelation between designing GetFace and its testing. In particular the control structure, with its explicit goals, permits testing of additional experts in situ. We are thus able to determine directly the competencies of new experts and their interaction with the rest of the system. For other reasons we are currently building software which allows controlled distortions of normal face images, both in terms of misplacing features and also by generating novel views of the face. We hope to use these faces in the future to investigate the tolerance of GetFace to unusual input. It is known that the human face processing system slows down significantly when presented with unusual views. We would certainly not aim for a greater tolerance from GetFace.

REFERENCES

Asada, H., & Brady, M. (1986). The curvature primal sketch. *IEEE Transactions on Pattern Analysis and Machine Intelligence, 8(1)*, 2–14.

Baron, R.J. (1981). Mechanisms of human facial recognition. *International Journal of Man-Machine Studies, 15*,137–178.

Binford, T.O. (1982). Survey of model-based image analysis systems. *International Journal of Robotics Research, 1(1)*, 18–64.

Baldock, R., & Towers, S. (1988). First steps towards a blackboard controlled system for matching image and model in the presence of noise and distortion. In J. Kittler (Ed.), *Pattern Recognition, 4th International Conference* (pp. 429–438). Berlin & Heidelberg: Springer-Verlag.

Bruce, V., & Young, A.W. (1986). Understanding face recognition. *British Journal of Psychology, 77*, 305–327.

Canny, J. (1986). A computational approach to edge detection. *IEEE: Transactions on Pattern Analysis and Machine Intelligence, 8(6)*, 679–698.

Craw, I., Ellis, H.D., & Lishman, J.R. (1987). Automatic extraction of face-features. *Pattern Recognition Letters, 5(2)*, 183–187.

Draper, B.A., Collins, R.T., Brolio, J., Hanson, A.R., & Riseman, E.M. (1988). Issues in the development of a blackboard-based schema system for image understanding. *International Journal of Computer Vision, 2*, 189–218.

Erman, L.D., Hayes-Roth, F., Lessor, V.R., & Reddy, D.R. (1980). The hearsay-II speech-understanding system: Integrating knowledge to resolve uncertainty. *ACM Computing Surveys, 12*, 213–253.

Engelmore, R., & Morgan, T. (1988). *Blackboard systems.* Wokingham, England: Addison-Wesley.

Kanade, T. (1977). Computer recognition of human faces. *Interdisciplinary Systems Research* (Vol. 47). Basel, Switzerland, & Stuttgart, Germany: Birkh.

Kelly, M.D. (1971). Edge detection in pictures by computer using planning. In B. Meltzer

& D. Michie (Eds.), *Machine intelligence 6* (pp. 397–409). Edinburgh, Scotland: Edinburgh University Press.

Klinger, A. (1984). Multiresolution processing. In A. Rosenfeld (Ed.), *Multiresolution image processing and analysis* (pp. 77–85). Berlin & Heidelberg: Springer-Verlag.

Marr, D. (1982). *Vision.* San Francisco, CA: Freeman.

Niemann, H., Bunke, H., Hofmann, I., Sanger, G., Wolf, F., & Feistel, F. (1985). A knowledge based system for analysis of gated blood pool studies. *IEEE: Transactions on Pattern Analysis and Machine Intelligence, 7(3),* 246–259.

Nazif, A.M., & Levine, M.D. (1984). Low level image segmentation: An expert system. *IEEE:T-PAMI, 6(5),* 281–303.

Nagao, M., Matsuyama, T., & Mori, H. (1979). Structural analysis of complex aerial photographs. *Proceedings of the Sixth International Joint Conference on Artificial Intelligence (IJCAI-79)* (pp. 610–616). Los Altos, CA: Morgan Kaufman.

Pratt, W. K. (1978). *Digital image processing.* New York: Wiley.

Sakai, T., Nagao, M., & Fujibayashi, S. (1969). Line extraction and pattern detection in a photograph. *Pattern Recognition, 1,* 233–248.

Sakai, T., Nagao, M., & Kanade, T. (1972). Computer analysis and classification of photographs of human faces. *Proceedings of the First USA-Japan Computer Conference* (pp. 55–62).

Stonham, T. J. (1986). Practical face recogniton and verification with WISARD. In H. Ellis, M. Jeeves, F. Newcome, & A. Young (Eds.), *Aspects of face processing* (pp. 426–441). Dordrecht, Netherlands: Martinus Nijhoff.

Towers, S. (1987). *Frames as data structures for SBS* (Tech. Rep.). Edinburgh, Scotland: Pattern Recognition and Automation Section, MRC Clinical and Population Cytogenetics Unit, Western General Hospital.

2
Mouth Shapes and Speech*

N. Michael Brooke
School of Mathematical Sciences
University of Bath
Bath, England

INTRODUCTION

Speech signals are conventionally treated as being purely acoustical, and there is now a vast literature in the field of acoustical speech processing. Indeed, much of the contemporary research effort is aimed at the production and application of practical devices for the automatic synthesis and recognition of acoustical speech signals. Nonetheless, there exists a second, visual, component in the production and perception of speech due to the visible movements of the mouth, face, and head. This component forms the basis of lip- (or speech-) reading. It has long been known that completely deaf observers can obtain useful linguistic information by looking at the faces of speakers. If the deaf were the only category of observers who could benefit from the presentation of visual information, lipreading would be an interesting skill worthy of study, but not one necessarily associated with the study of acoustical speech production and perception. However, there is a tradition for the descriptions of the different processes in the speech chain to come together, and there are at least two indications that the visual aspects of the speech signal can be relevant to studies in the acoustical domain.

First, speech intelligibility is improved if the face of the speaker can be seen whenever there is degradation of the acoustical signal either by noise or by hearing impairment (Erber, 1975). It has been shown (MacLeod & Summerfield,

---

* Part of the author's work described in this chapter was supported by research grants from the Medical Research Council and the Science and Engineering Research Council, and was carried out in collaboration with Dr. A. Q. Summerfield at the MRC Institute of Hearing Research, Nottingham, by permission of the Director, Prof. M. P. Haggard. The work on automatic visual speech recognition was largely carried out under a consultancy in collaboration with Dr. E. D. Petajan at AT&T Bell Laboratories in Murray Hill, New Jersey, by permission of the management at AT&T.

1987) that, when speech is presented in noise, the benefit gained by seeing a talker's face is, on average, equivalent to a gain of 8–10 dB in the signal-to-noise ratio. The perception of speech in this bimodal domain is commonly exploited by normal-hearing subjects as well as by hearing-impaired subjects, for whom it is central to everyday communication. The processes by which observers acquire and process visual information and integrate it with auditory information when they perceive speech in this bimodal domain are not well understood, but are fundamental to any comprehensive theory of speech perception (Summerfield, 1987). That the modes do interact is demonstrated by experiments in which observers presented with conflicting auditory and visual stimuli may perceive speech events which are a fusion of the two separate stimuli (McGurk & Mac-Donald, 1976). Thus, for example, a visual /ga/ and an acoustic /ba/ is commonly perceived as a /da/. Secondly, in addition to facial gestures such as the movements of the eyebrows, which can convey linguistic cues without being directly involved in the production of speech sounds (Ekman, 1979), the gestures in the oral region in particular are directly related to modulations of the vocal tract configuration which affect its length and shape and hence its acoustical output. Consequently, the visible gestures which are significant in lipreading are also relevant to the articulatory movements and acoustics of speech production. In particular, visual cues can indicate the place of articulation of consonants. This information is frequently difficult to extract from acoustical signals, because the auditory cues tend to be associated with low-intensity, short-duration details in the mid- to high-frequency regions of the acoustic spectrum whose detection requires good frequency resolution. These are precisely the cues which are most susceptible to noise interference. In contrast, the nonvisible articulators like the larynx and velum tend to control aspects of the speech signal, such as the voicing and nasality of consonants and the stressing and intonation of speech, which are more resistant to corruption by noise because they usually correspond to more intense and slowly changing patterns in the low-frequency part of the acoustic spectrum. There is thus a useful degree of complementarity between the auditory and visual speech cues.

This chapter will focus on two contemporary areas of study concerning visual speech cues and will indicate how they depend upon the measurement and analysis of the visible speech articulations. The first is basic scientific research to investigate analytically the mechanisms of visual and audiovisual speech perception. A clearer description of facial speech movements and of their perception, especially in relation to hearing impairment, would not only be valuable in itself but would serve to strengthen programs for the rehabilitation of the hearing impaired by suggesting scientifically well-founded improvements to existing techniques for assessing and teaching speech-reading skills. While the problem of providing controlled auditory stimuli has been overcome by the development of synthesizers whose acoustical output can be prescribed, one of the major difficulties in devising analytical experiments in visual and audiovisual speech perception is the provision of reproducible visual stimuli which can be completely specified a priori and can be controlled so as to produce, for example, graded continua of articulatory gestures. Films and videorecordings of speakers may therefore be unsuitable, because human talkers cannot accurately reproduce, let alone controllably vary, their gestures. Animated computer graphics displays of facial images simulating speech articulations, on the other hand, can fulfil all of these requirements. In addition, unlike human faces, they can be readily modified to generate ranges of visual stimuli, including entirely novel, artificial ones. The second area of relevance is applied research which is aimed at exploiting visual speech cues to enhance automatic speech recognition, especially in noisy conditions. Unless the target vocabulary is very small, the performance of the current generation of conventional, acoustic speech recognizers is significantly degraded even by moderate levels of background noise. In accordance with reasons that have already been outlined, it has been found that useful phonetic information can be obtained by lipreading alone (Finn, 1986). There is therefore good reason to suppose that the performance of automatic speech recognition systems could be improved if the input of acoustical information were augmented by a suitable form of input representing the visual speech cues, especially those arising in the oral region.

Whether it is desired to synthesize visual stimuli for speech perception experiments or to analyze visible speech articulations for recognition systems, it is necessary to record, measure, and analyze the facial gestures of talkers. Graphical syntheses rely upon suitable methods for describing the kinematics of speech articulations. Analyses of human gestures can be used (a) to direct the overall design of a computer graphics model of a talking head by suggesting the identity of an adequate, minimal set of key facial features and the nature of their mobility; (b) to animate a quasicinematographical computer-graphics display at a low level by describing the time-varying positions, frame by frame, of the set of key features; and (c) to develop higher levels of kinematic control for graphical syntheses by providing the data from which the generation of movements by rule may be derived (for example, Fowler, Rubin, Remez, & Turvey, 1978; Browman & Goldstein, 1985). Visual speech recognizers depend upon the capture of speech gestures from sequences of images of talkers' faces. The images can be processed to extract essential visual cues or features. Orthodox recognition then proceeds by matching the time-varying templates of features taken from a test utterance against the stored templates of utterances in the recognizer's vocabulary.

MEASUREMENT OF FACIAL SPEECH ARTICULATIONS

The recording and measurement of facial movements is a difficult problem, for a number of reasons. First, the quantitative estimation of the movements of specific facial features requires a reference frame to be set up within which their time-varying positions can be fixed. Because the face is a complex and highly mobile object whose detailed topographical configuration varies widely between individuals, it is difficult to set up a canonical reference frame. Secondly, in natural speech production, linguistically informative global movements of the head and body accompany, and are superimposed upon, the purely articulatory movements which are necessary in order to generate the acoustical speech signal itself (Ekman, 1979). Rigorous analysis of articulatory movements therefore requires their separation from the global movements, and two alternative solutions are possible. In one, the head can be firmly constrained to a fixed position by clamps and supports. In this case, the reference frame can readily be defined externally to the head. Unfortunately, this method may be stressful to speakers and lead to unnatural speech production. The alternative approach is to allow speakers freedom of movement. In this case, the reference frame must be set up internally to the speakers' heads, and the analysis procedures become more complex in consequence. The third problem in measuring articulatory movements is that they are rather small, and the measurement system must therefore be sensitive. Articulatory excursions from a neutral facial position, in which the lips and jaw are lightly closed, rarely if ever exceed 25 mm. Fourthly, articulatory movements associated with a specific speech utterance may vary widely between speakers and, even for a single speaker, over time (Montgomery & Jackson, 1983). In addition, the articulatory configuration for a single phonetic event may be modified by its phonetic context (Benguerel & Pichora-Fuller, 1982). It is therefore difficult to define comprehensive corpora of utterances for measurement. A fifth problem is that most measurement systems are to a degree invasive, so that measurements are always liable to represent somewhat nonnatural production processes.

Point Measurements

In general, studies of articulatory movements have concentrated on the internal vocal tract, because this domain is relevant to acoustical speech analysis and synthesis. A very large number of methods have been employed and have been surveyed elsewhere (Brooke, in press). They include ultrasound imaging, electropalatography, and magnetometry. The most widely used technique, however, probably remains x-radiography, including X-ray microbeam methods (for example, Perkell, 1969; Fujimura, Kiritani, & Ishida, 1973; Kiritani, Itoh, & Fujimura, 1975). A large body of data has now been amassed (for example, Perkell, 1969; Fant, 1960). While these studies have yielded useful data about the motions of points within the vocal tract, they do not supply detailed information about movements of the front of the face. The only relevant data provided by measurements of the internal tract, which are conventionally made in the midsagittal plane, are those relating to the vertical separations of the centers of the inner lip margins and of the teeth. Even the observed movements of the tongue are not directly related to their visible projection in the frontal plane.

Studies of the facial articulations, particularly those around the lips and jaw, have most commonly been made by optical methods, although strain-gauge measurements have also been used (for example, Muller & Abbs, 1979; Nelson, Perkell, & Westbury, 1984; Kuehn, Reich, & Jordan, 1980). Talkers have generally been recorded on cinefilm, and measurements have been made, frame by frame, of the movements of individual points (for example, Lindblom & Soron, 1965; Fujimura, 1961). In some of the experiments, light-reflecting beads or light-emitting diodes were used to mark the positions of sets of key articulatory points so that they could each be tracked automatically from frame to frame (McCutcheon, Fletcher, & Hasegawa, 1977; Sonoda & Wanishi, 1982). One system (McCutcheon et al., 1977) used video-processing circuitry to find bright spots during the scanning of the frames by a video camera. In most experiments, the talker's head was fixed, so that an external reference frame could be used to measure the movements. Brooke and Summerfield, however, allowed the talker freedom of head movement in their experiments (Brooke & Summerfield, 1983). The talker was positioned adjacent to a mirror angled at 45 degrees to the camera's principal axis so that the head could be simultaneously recorded in the front and side views. Recordings were made on videotape, using a shuttered camera to improve the resolution of the images, particularly during rapid consonantal articulations. An analogue storage disc was used to recover the recordings one frame at a time. The output from the disc was fed to a television monitor via a video-cursor device. This device imposed a pair of 'crosswires' in the plane of the image which could be moved under control of a joystick. On command, the digitized coordinates of the crosswire intersection could be transmitted to a microprocessor for storage. It was therefore possible to locate and record the positions of each of a set of 13 marked points around the inner and outer lip margins and jaw of a talker for each of a sequence of frames representing a specific speech utterance. Four additional points on the head were marked and tracked. They were chosen to lie at locations which were not involved in the articulatory movements of speech production. By tracking these four "skeletally rigid" points from frame to frame, it was possible to determine the orientation of the head. Standard mathematical techniques were then applied to separate the global head and body movements from the articulatory movements. The calculated articulatory movements were then plotted as time-varying trajectories in each of the x-, y-, and z- directions for each measured point. The movements were plotted as displacements from a neutral facial position, and examples are shown in Figure 2.1.

Since the movements of individual points in a specific direction are single-valued functions of time, they are relatively easy to manipulate. For this reason they are commonly used to construct and analyze models of speech articulation. They are also the primary driving data for animated computer-graphics syntheses of talking faces, because (a) they define the positions of the ends of the vectors or of the vertices of the polygonal shapes from ensembles of which the displays are

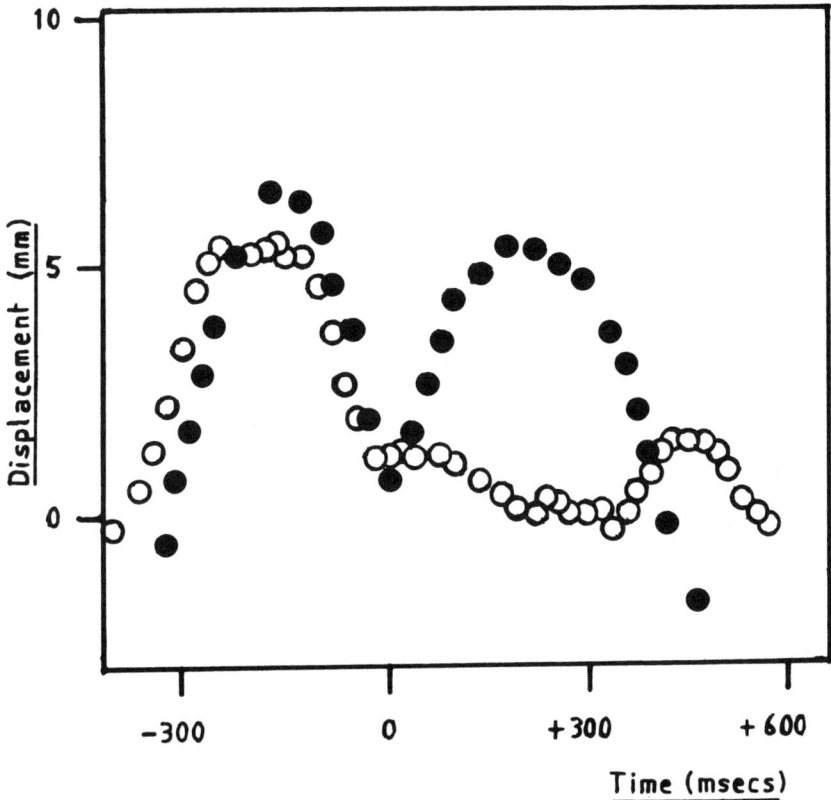

Figure 2.1. Examples of measured articulatory trajectories showing the vertical movements of the corner of the outer lip margin during productions of the syllables /aba/ (filled circles) and /abu/ (open circles). Displacements are measured downwards from a neutral facial position, and the time origin is located at the medial bilabial closure.

constructed as a series of time-varying frames, and (b) they can be simply processed in both the temporal and spatial dimensions to produce variations in the kinematic characteristics of the displays. The major disadvantage of point measurements is that they fail to capture many aspects of the shape of a feature like the mouth, or of its movements, as a whole.

Whole-mouth Measurements

With the development of video technology, it has now become possible to capture and store digitized representations of single frames of a video display, in which the images are stored as two-dimensional arrays of pixels. Each pixel holds a digital value which encodes the intensity level of a monochrome image, or the hue and intensity of a color image. It is then possible to access individual pixels or subgroups of pixels and process them by computer. Thus, for example, it is possible to take all, or a 'windowed' part of, an image and perform operations (a) to threshold the image in order to highlight particular regions and eliminate others; (b) to rescale the color, intensity, or spatial resolution of the image; (c) to smooth out minor irregularities of colour or intensity; and (d) to detect the edges of regions of a particular intensity or color. The raw and processed images can also be held in computer storage, either directly, as pixel maps, or in the form of a compacted code. A very wide range of image-processing operations is available (Ballard & Brown, 1982).

Digital image-processing methods are therefore appropriate, and have been applied, to the study of visible speech articulations in which the time-varying visual patterns of the whole oral region are relevant (for example, Nishida, 1986; Petajan, 1984; Petajan, Bischoff, Bodoff, & Brooke, 1988). Nishida used image-processing methods to determine the degree of difference between successive images in a sequence. It was then possible, for example, to find the start and end of speech utterances by finding the points at which the mouth was at rest and the images were therefore not changing significantly. Petajan used a miniature monochrome television camera mounted on a head boom to eliminate most of the effects of the global head and body movements. The camera returned video data at 60 frames per second, corresponding to a digital input data rate of about 3.5 megabytes per second. Special-purpose real-time image-processing hardware was developed and used to compress the image data. The hardware windowed the images so that the oral region could be isolated. It then applied a single threshold to the video signal so that binary, black-and-white images were generated from the grey-scale input. The threshold was set interactively so that the nostrils always appeared as dark areas and the mouth region contained no dark areas when the mouth was closed. The binary images were then smoothed along each raster line to eliminate very short runs of black or white pixels resulting from jitter in the video signal. Finally, the hardware encoded the smoothed binary images using a form of predictive differential quantization, details of which have been given elsewhere (Petajan, 1984). The compacted data, in the form of a contour code which occupied about 2 kilobytes per frame, were transferred to the store of a computer system for further processing. The stored contour code for each image was used to reconstruct a series of regions, or areas of constant grey-level lying within a closed boundary. Figure 2.2 shows examples of the kind of image sequences which were generated this way by Petajan's system.

The area, perimeter, and extreme horizontal and vertical coordinates of each region were used as parameters to identify it. In order to eliminate any residual effects of head and body movements not dealt with by mounting the camera on a head boom, the regions identifying the nostrils were tracked from frame to frame as follows. A single reference frame was chosen in which the nostril regions

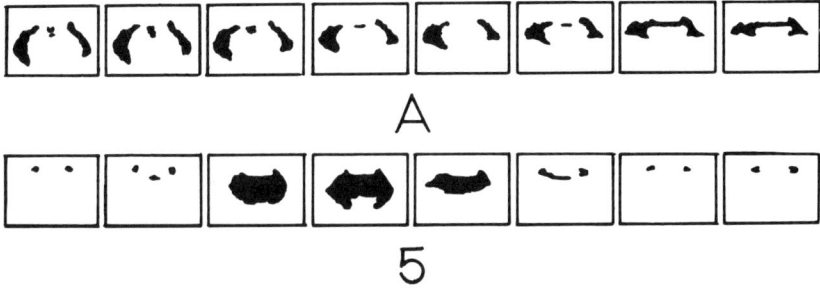

Figure 2.2. Examples of frames from image sequences captured with the mouth-imaging system of Petajan, described in the text. They show the changing oral shapes produced by a speaker enunciating (a) the letter word "A" (upper row) and (b) the digit word "five" (lower row). The sequences run from left to right and show frames sampled at 67 msec. intervals. Note that the teeth and tongue tip appear as white areas within the dark oval cavity.

were manually identified and labelled. Two software windows were then set up in a fixed relative configuration so that the first, or nostril window, enclosed the nostril regions and the second, or mouth window, was large enough to enclose all mouth regions which were likely to be encountered during speech articulations. The mouth was found to remain in a virtually constant position relative to the nostrils during speech production. The first frame from each measured image sequence was then processed by finding the distances in the region parameter space which separated every region from the nostril regions of the labelled reference frame. The distances were ranked, and the two regions which were closest to the nostril regions in the reference frame were labeled as nostrils (they had also to satisfy other geometrical constraints, such as separation). The two software windows were then shifted by the same distance in the same direction so as to recenter the detected nostril regions in the nostril window. Subsequent frames were processed in a similar way, except that the continuity of head movements was exploited and the region matching carried out only within the nostril window set for the preceding frame.

Following nostril tracking, all low-intensity regions falling within the mouth window were classified as mouth regions. In early experiments (Petajan, 1984), the total area and perimeter, plus the extreme horizontal and vertical coordinates, accumulated over all mouth regions, were used as parameters to represent essential features characteristic of the oral configuration in a single image. By plotting each of these parameters as a function of time for a sequence of images, templates could be formed for pattern matching (see below). In later studies (Petajan, Bischoff, Bodoff, & Brooke, 1988; Petajan, Brooke, Bischoff, & Bodoff, 1988) attempts were made to utilize the oral regions directly for pattern matching, so as to preserve as much of the visual information as possible. The core of conventional pattern-matching techniques is the derivation of a suitable metric for estimating quantitatively the separation between patterns in the pattern space. For

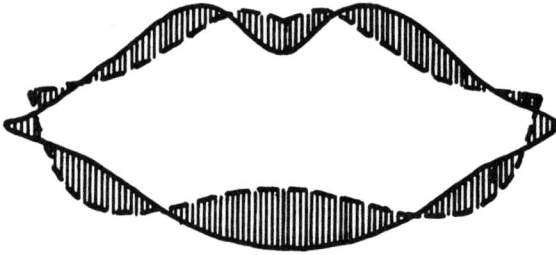

Figure 2.3. Two illustrative mouth shapes in which the outlines mark the boundaries of the low-intensity regions of the binary images which correspond to the oral cavity. The shaded areas show how the "logical exclusive or" of corresponding pixels represents the degree of nonoverlap of the image pair.

binary, black-and-white images such as those produced in the mouth window by Petajan's system, a computationally efficient method has been suggested which is sensitive to the shape and size of the mouth regions. The distance in pattern space (D) between two images is defined in terms of the total area of the logical exclusive or formed between the mouth regions of the image pair (X) and the total area of the mouth regions in each of the two images (A and B):

$$D = X/(A + B)$$

Figure 2.3 illustrates how the metric represents the degree of nonoverlap between the mouth regions in two images. D is a bounded and normalized metric, that is, $0 <= D <= 1$. If $A = B = 0$, D is set to 0. If one of A or B is zero, D is set to 1. This is justified, since low-intensity mouth regions tend to disappear only when there are silences or bilabial closures. Even a small mouth opening is therefore likely to represent a perceptually and phonetically different speech event. This metric has also been applied to visual speech recognition experiments, as described below.

As Figure 2.2 shows, the teeth and tongue tip can be visible in the images obtained with Petajan's system and are known to be perceptually significant (McGrath, Summerfield, & Brooke, 1984). However, in one experiment (Brooke & Petajan, 1986) the teeth were blacked out so that fuller studies of the shapes of the inner lip margins which bounded the mouth region could be carried out. The primary objective was to test a method for measuring differences in lip shapes. This measure was then applied to studies of coarticulation in the visible domain both within and between speakers. The characteristic shapes of the lips were represented by a radial function. A coordinate system was defined in which the origin was the center of a line joining the lip corners, which were readily identifiable. The radial distance of the lip margin from the origin was then plotted at 60 equal intervals as the radius vector swept a circle. The zero angle

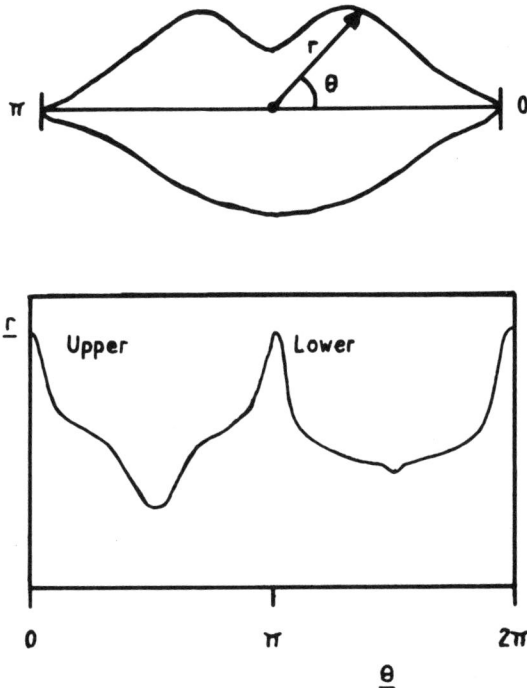

Figure 2.4. A schematic diagram to show how the radial function described in the text can be used to represent the shape of the inner lip margins.

was taken to be the left corner of the lips. Figure 2.4 shows a typical radial function.

The radial function has a number of advantages. It is a normalized representation, so that lip shapes can be compared independent of the length of the lip perimeter; it can be manipulated to correct for rotations of the head by shifting the plot along the angular axis; it can be manipulated to scale lip shapes to a standard mouth size, by shifting the plot in the radial axis and it is easy to compute from the contour-coded image data. In addition, the radial function clearly delineates the upper- and lower-lip margins and is sensitive to changes in lip shape. The special significance of the radial function, however, was that it became possible to define a simple metric to estimate the similarity of lip shapes in pairs of images, irrespective of the sizes of the oral areas. For two frames, i and j, the metric is defined by:

$$0 <= S_{ij} = [\Sigma \ r_i \cdot r_j]^2 \ / \ [\Sigma \ r_i^2 \cdot \Sigma \ r_j^2] <= 1$$

where r_i and r_j are the radial values at corresponding angles for the two frames and are summed over all angular intervals. The similarity metric is effectively the normalized sum of squares of deviations after least-squares fitting one of the radial functions to the other so as to maximize their overlap. The results of the studies verified the usefulness of the radial function and similarity metric and have been fully reported elsewhere (Brooke & Petajan, 1986).

While mouth shapes derived from image processing are usually applied to visual speech recognition experiments because they encapsulate features of complete images of the primary facial articulators, they have also been used as tokens to drive a prototypical visual speech synthesiser (Storey & Roberts, 1987). In this experiment, representative single oral shapes captured from binary images of speakers enunciating a selection of vowels and consonants were stored as pixel maps. Animated visual syntheses were then generated from phonetic transcriptions of utterances by concatenating and displaying appropriate sequences of the token images. The syntheses were generated at frame rates very much lower than 25 frames per second and did not include adjustments for variations in the mouth shapes due to coarticulatory effects.

APPLICATIONS OF MOUTH MOVEMENT MEASUREMENTS

Video Speech Synthesisers

A number of computer-graphics syntheses of facial speech movements have been implemented. They have been fully reviewed elsewhere (Brooke, in press). They fall into two main types: vector displays and raster displays.

Raster graphics are capable of producing very high-quality, detailed displays, including texturing and shading of complex surfaces. Facial images are usually generated by building the face model from a large number (200 or more) of polygonal elements which can then be smoothed and shaded to generate the final image. By controlling the position of the vertices of the polygons, it is possible to generate sequences of images which can be filmed one frame at a time and subsequently screened to simulate real-time articulatory movements (for example, Parke, 1975, 1982). Models of this kind have been automatically synchronized with speech by selecting specific configurations of the polygonal elements to match the acoustic features of the speech signal at each of a succession of points in time through the utterances (Lewis & Parke, 1987). The vertices of the polygonal elements may be based on the underlying anatomical structure by supposing them to represent a network of points on the skin surface. The skin can be viewed as an elastic sheet whose movements are controlled by the action of sets of muscles. The action of a muscle at one point in the network can then be simulated by computing the displacements which are propagated outwards from that point throughout the network (Platt & Badler, 1981). Polygonal and network models, however, suffer from a number of drawbacks as a source of optical stimuli in speech perception experiments. First, they are computationally expensive and time-consuming to generate, and require relatively sophisticated graphics displays. Secondly, it is difficult to define and supply the large amount of detailed data upon which they would depend for the accurate generation of articulatory movements, particularly in the oral region. Thirdly, once defined, it is difficult to modify the facial models so as to generate differing facial topographies, which can be important in perception experiments, as will be discussed below.

Vector displays, on the other hand, while they can generate only outline diagrams and are therefore unable to model surfaces like the tongue, or skin texturing and shadowing such as occurs when the cheeks are puffed out or the lips are puckered, nonetheless have a number of compensating advantages. Principally, they are cheap and relatively easy to generate on low-cost displays. One early form of vector synthesis (Montgomery, 1980; Montgomery & Soo Hoo, 1982) used a fixed, preordered set of 143 vectors to model the lips and upper teeth of a talker. Real-time animation was achieved by supplying complete specifications of the vector set 30–40 times per second to a graphics terminal. Tracings from a videotape recording of a speaker were used to define a series of vector sets for idealized target phonemes. Speech simulations were achieved by concatenating a sequence of vector sets for the utterance to be simulated and interpolating additional vector sets. Anticipatory and perseverative coarticulation were empirically modeled by modifying the target shapes according to the influence of preceding and following phonemes. The chief disadvantages of this model were that (a) the separate articulators were not independently defined, so that they could not be individually manipulated to produce certain kinds of artificial visual continua appropriate to perceptual studies; (b) the vector set would have to be completely redefined if augmented or reduced sets of visible articulators were to be modeled and displayed; and (c) not all phonemes can be modeled by single target frames—some may be signaled by different kinds and rates of movement.

The vector graphics simulation developed by Brooke attempted to overcome many of these disadvantages. It has been described in greater detail elsewhere (Brooke, 1982, 1988, in press, 1989; Brooke & Summerfield, 1983; Brooke, McGrath, & Summerfield, 1984). It was designed to permit the independent specification of (a) the topography, or set of facial features to be drawn by the display; (b) the speaker identity, or specific facial shape to be drawn with a given topography; and (c) the utterance to be simulated. By varying the topography it was possible, for example, to draw a picture of an essentially complete face or a more detailed picture of a small region of the face, such as the oral region. The specification was provided as an ordered sequence of lines and circular arcs referred to a set of key points. The movement permitted to each key point was also prescribed. Thus points could be (a) fixed, (b) independently variable (that is, they move according to the utterance being simulated), or (c) dependently variable (that is, their movement is related to the movement of one or more of the independently moveable key points). The last category dealt with points like the angle of the jaw, whose movement is physically constrained through the rigid

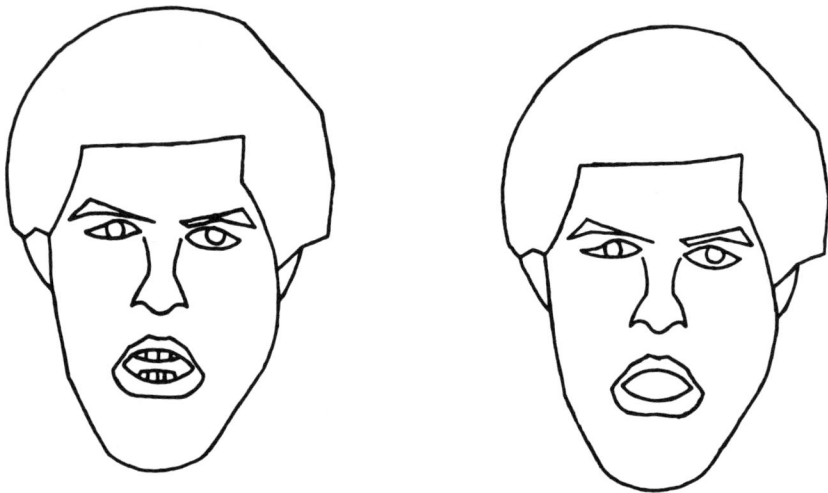

Figure 2.5. Examples of two frames generated by Brooke's video speech synthesizer. They both represent the vowel nucleus in a resynthesis of the syllable /bib/ and illustrate the use of the synthesizer to generate different facial topographies in which the teeth are (a) present and (b) absent.

mandible to the movement of the base of the jaw. Facial features like the teeth, which are only partially or intermittently visible, could be built into the model by defining the topographical features in closed polygons at different levels of visibility, or planes in the z-axis. A hidden-line removal algorithm was subsequently used to remove occluded features from the final display. The speaker identity was determined by specifying the spatial positions of each of the key points for a neutral facial configuration (see "Point measurements"). A normalized coordinate system was adopted so that display generation would not be dependent upon a specific graphics device. Utterance simulation was driven by supplying an articulatory trajectory (see "Point measurements") for each direction of movement at each of the independently moveable key points. It was therefore possible to compute the position of each of the key points in the diagram for a series of frames at closely spaced time intervals (1/25 to 1/50 second). Finally, the lines and arcs connecting the key points in the diagram according to the topographical specification could be computed for each frame and stored as sets of vectors. The animated speech simulations were accomplished by recalling the stored vector sets, converting them to device-specific coordinates, and streaming them to the screen of a graphics terminal. In a typical application, up to 10,000 vectors per second could be generated. It is possible to handle this density of graphical data on a relatively inexpensive, unbuffered display device. Figure 2.5 shows typical frames from a facial display and shows how the package can be used to generate displays with varying topographies.

This approach to synthesis has three main virtues. First, by separating the data which specify the model into categories, it is possible to alter some attributes of the visual displays without changing others. For example, the same face can simulate different utterances, or identical utterances can be simulated by different faces. Also, because the topographical specification identifies the articulators individually, they can be modified very easily to present varying types of visual stimuli. For example, the jaw could be omitted from the display, or additional features like skin folds could be added. Secondly, complex movements can be generated through the specification of articulatory gestures at a small number of key points. Thus the movements of the entire jaw and cheek outlines are essentially governed by the movement of one point at the base of the jaw. This greatly reduces the overheads of data acquisition and storage. Thirdly, since facial topographies, speaker identities, and utterance specifications are provided through external data, these may themselves be independently generated and manipulated. In particular, the articulatory trajectories which define the utterances may be manipulated in both the spatial and temporal dimensions, so as to produce the graded continua of stimuli which are so frequently needed in perceptual investigations. It is also easy to generate entirely artificial stimuli, for example, by decoupling the movements of articulators like the lower lip and the jaw, whose movements are normally constrained physically and dynamically so as to be broadly similar in both direction and magnitude. Thus it is possible to produce continua in which, at one extreme, the jaw remains static when the lower lip is depressed, while at the other extreme, the lips are stationary despite the lowering of the jaw.

Perceptual Experiments With A Video Speech Synthesizer

Brooke's video speech synthesizer has been successfully applied to a number of analytical experiments in visual and audiovisual speech perception, the results of which have been fully reported and discussed elsewhere (Brooke et al., 1984; McGrath et al., 1984; McGrath, 1985; Summerfield, MacLeod, McGrath, & Brooke, 1989). In these experiments, experimental articulatory trajectories were derived from measurements at 13 points around the lip margins and jaw of a talker enunciating 11 British-English, nondiphthongal vowels in a /bVb/ context. The articulatory trajectories were used directly to resynthesize a real-time animated display of the lips, jaw, and cheek outlines of a face generated with the video speech synthesizer. The experiments were undertaken (a) to confirm that the teeth play a role in distinguishing between the open vowels and between rounded and unrounded vowels, and (b) to verify that vowels produced by a synthesiser were perceived in the same way as the natural vowels from which they were resynthesized. The flexibility of the synthesiser was exploited in the first experiment to generate facial images in which the teeth were (a) present and (b) absent, but in which all other features and gestures were identical. It confirmed the relevance of the teeth to visual cueing. In the second experiment, vowel identification performance by observers was compared using (a) the two types of synthetic image and (b) videorecordings of the human talker's face in which (i) the whole face was visible, (ii) only the lips and teeth of the talker were visible, and (iii) only the lips of the talker were visible. Statistical analysis of the results under each condition by multidimensional scaling indicated that observers were treating the synthetic stimuli in much the same fashion as the natural stimuli, and that the synthetic stimuli, in which only the lips, teeth, and jaw were mobile, were conveying virtually as much information as the natural stimuli in which only the lips and teeth were visible. Further evidence that the synthetic stimuli could be perceived as speech-like was obtained by combining conflicting auditory and synthetic visual stimuli and evoking the McGurk effect (see "Introduction"). Despite its limitations, the experiments indicated the usefulness of the video synthesizer as a tool in perceptual experiments.

Automatic Visual Speech Recognition

Early experiments in speaker-dependent, automatic visual speech recognition were carried out on an isolated-word vocabulary composed of the spoken digits from 'zero' to 'nine' (Petajan, 1984). Test utterances were matched against each of the vocabulary entries in turn by comparing sets of templates consisting of the time-varying parameters such as mouth areas, perimeters, and extreme coordinates whose derivation was described in "Whole Mouth Measurements." The vocabulary entry whose set of reference templates most closely matched the test template set was chosen as the best match. Dynamic time warping was employed to align the templates. These experiments showed that correct recognition rates of over 95 percent could be attained, although the templates embodied only limited aspects of the mouth shapes.

More recent experiments used vocabularies of both digit and letter words (Petajan, Bischoff, Bodoff, & Brooke, 1988; Petajan, Brooke, Bischoff, & Bodoff, 1988). The letter vocabulary included words which were acoustically and visually similar. In an attempt to maximise the use of the available visual information, pattern matching of test and vocabulary entries was accomplished by comparing sequences of images directly. The test sequence was compared to each of the vocabulary sequences in turn. The distance between each pair of image sequences was computed by using the D metric described in "Whole Mouth Measurements" to find the separation between corresponding pairs of images in the two sequences and then summing the separations over all frames, having aligned the two sequences by dynamic time warping, as in the earlier experiments. The vocabulary entry which led to the smallest cumulative distance value was selected as the best match. The correct recognition rate for the digit words varied between 95 and 100 percent for the four speakers tested. That for the letter words varied between 72 and 80 percent. However, these figures did not allow for confusions between letter words whose articulations are visually indistinguishable. Thus, for example, "A" and "K" differ only in the unseen

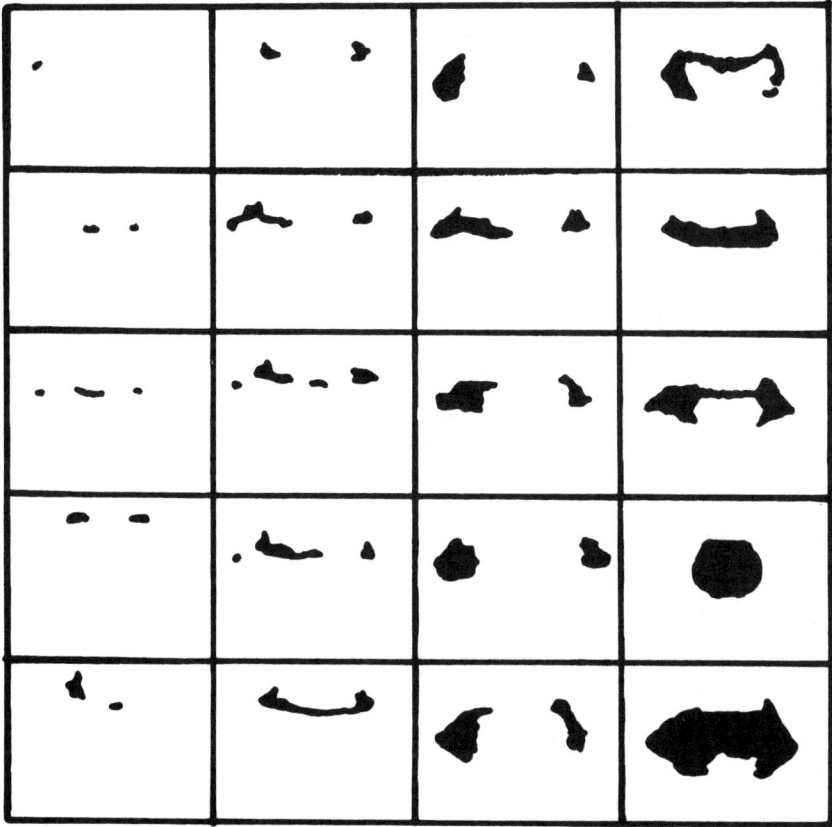

Figure 2.6. Examples of 20 entries from a codebook of approximately 256 oral shapes constructed for a single speaker by clustering, as described in the text.

glottal stop; "B" and "P" only in the nonvisible voicing of the initial plosive; and so on. Other visually indistinguishable pairs are "C" and "Z" (in American English!), "D" and "T," "S" and "X", and "Q" and "U". When allowances were made for confusions within these pairs, the correct recognition rates varied between 83 and 94 percent for the four speakers.

A variation upon this experiment was also carried out in which the images were 'vector quantized.' A special image sequence was formed by concatenating visual tokens of all the letter and digit words. The D metric was used to separate all of the images in the concatenated sequence into a series of clusters. Clusters were defined such that a new one was formed only when an image failed to lie within a threshold distance of every member of an existing cluster. The procedure was iterative and adjusted the threshold distance until approximately 256 clusters were obtained. A codebook was then formed in which each cluster was represented by one image. Part of such a codebook is shown in Figure 2.6.

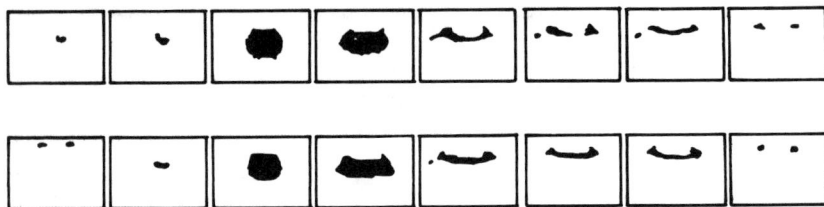

Figure 2.7. Selected frames from an image sequence showing oral shapes produced by a speaker enunciating the word "one." The sequences run from left to right and show frames sampled at 67 msec. intervals. The upper row shows the original images, and the lower row the vector quantized version in which each frame has been replaced by an entry from a codebook of mouth shapes, as described in the text.

In addition, a complete table of intercluster separations was computed. It was then possible to replace each image in both test and vocabulary image sequences by the image in the codebook which best matched it. Figure 2.7 shows raw images which have been vector quantized in this way.

The matching of pairs of image sequences was achieved by the computationally efficient expedient of looking up the stored tables of intercluster distances to find the separation between image pairs and summing the separations over all the frames in the sequences. Correct visual recognition rates remained high, falling no more than 6 percent from the rates obtained without vector quantization for any of the four speakers in either digit or letter vocabularies.

While visual speech recognition experiments have thus far been conducted only on individual speakers with isolated words from small vocabularies, the results suggest that significant information can be derived even from those limited aspects of the oral gestures which can be mapped into a relatively small codebook of shapes.

SUMMARY AND CONCLUSIONS

Despite the difficulties that have to be overcome in order to measure and analyze accurately the movements of facial articulators during speech production, methods have been devised which can capture useful information both about individual points around the lips and jaw of speakers and about the varying shapes of the oral region as a whole. The results of observations of these kinds have been applied successfully (a) to the synthesis by computer graphics of animated facial displays which can simulate visible aspects of speech production, and (b) to the automatic, speaker-dependent recognition of single words which relies upon the visual cues alone.

Studies of synthesis and recognition are complementary. The application of visual speech synthesis to analytical investigations of visual speech perception has already shown, for example, that the teeth are perceptually significant features in the recognition of vowel productions (McGrath et al., 1984). This finding clearly suggests that successful visual speech recognizers would need to extract information about the teeth from images of a speaker's oral region, as Petajan's already does (Petajan, 1984). In the longer term, perceptual studies could lead towards an even more detailed description of the facial features and movements which would need to be extracted from facial images if the performance of automatic visual speech recognizers were to be maximized. In addition, studies of audiovisual perception may suggest how best to integrate acoustical and visual speech recognizers. The application of visual speech synthesis will therefore be likely to direct the future development of visual speech recognition.

Conversely, in order to extend the usefulness of synthesizers for experimental studies, their efficiency needs to be increased so that displays can be generated at rates which more closely approach real-time, and the quality needs to be refined so that even more lifelike images can be constructed. At present, for example, Brooke's synthesizer requires a timescale of approximately 50 to 200 times real-time to generate displays like those shown in Figure 2.5, even though the display itself can be screened in real-time, or less than real-time. Also, vector displays like Brooke's cannot accurately model features like the tongue which appear as textured areas rather than outlines. In particular, there is at present no unifying theory to account for the kinematics of speech production, so that it is difficult to estimate accurately either the number of parameters that are required to define an articulatory model of the face or to relate these parameters to underlying control mechanisms. The techniques for handling images in visual speech recognizers, however, offer potential tools for investigating a number of these problems. For example, the successful use of codebooks of oral shapes for visual speech recognition suggests that they can embody useful visual information within a bounded set of images. At present, no work has yet been completed to investigate the variation in performance of visual recognizers as the size of the codebook is changed, or to search for the underlying factors which discriminate between shapes in the codebook. Results from studies of this kind could be used in two ways. First, they could suggest how to simplify the image-processing algorithms while retaining a visual recognizer's optimal performance. Secondly, they could provide a sound scientific basis for the construction of highly efficient visual speech synthesizers which generated their displays by concatenating sequences of images selected from a minimal set of perceptually significant oral shapes (see Storey & Roberts, 1987). Other, longer-term applications of code books have also been envisaged (Petajan, Brooke, Bischoff, & Bodoff, 1988). Continuing work on visual speech recognition is therefore likely to influence strongly future developments in visual speech synthesis.

REFERENCES

Ballard, D.H., & Brown, C.M. (1982). *Computer vision.* Englewood Cliffs, NJ: Prentice-Hall.

Benguerel, A-P., & Pichora-Fuller, M.K. (1982). Coarticulation effects in lipreading. *Journal of Speech and Hearing Research, 25,* 600–607.

Brooke, N.M. (1982). Video speech synthesis for speech perception experiments. *Journal of the Acoustical Society of America, 71,* S77 (abstract).

Brooke, N.M. (1988). Looking at speech: Studying the analysis, recognition, synthesis and perception of visible speech gestures. In P. Salenieks (Ed.), *Computing: The next generation.* Chichester, England: Ellis Horwood.

Brooke, N.M. (in press). Computer graphics animations of speech production. *Advances in Computing and the Humanities.*

Brooke, N.M. (1989). Visible speech signals: Investigating their analysis synthesis and perception. In M.M. Taylor, F. Neel, & D. G. Bouwhuis (Eds.), *The structure of multimodal dialogue.* Amsterdam, Netherlands: North-Holland.

Brooke, N.M., McGrath, M., & Summerfield, A.Q. (1984). Visual speech perception experiments using a video speech synthesiser. *Journal of the Acoustical Society of America, 76,* S81 (abstract).

Brooke, N.M., & Petajan, E.D. (1986). Seeing speech: Investigations into the synthesis and recognition of visible articulatory movements using automatic image-processing and computer graphics. *Proceedings of the IEE Conference on Voice I/O Techniques and Applications (London)* (pp. 104–109). (IEE Conference Publication Number 258). London:IEE.

Brooke, N.M., & Summerfield, A.Q. (1983). Analysis, synthesis and perception of visible articulatory movements. *Journal of Phonetics, 11,* 63–76.

Browman, C.P., & Goldstein, L.M. (1985). Dynamic modelling of phonetic structure. In V. Fromkin (Ed.), *Phonetic linguistics.* New York: Academic Press.

Ekman, P. (1979). About brows: Emotional and conversational signals. In M. von Cranach, K. Foppa, W. Lepenies, & D. Ploog (Eds.), *Human ethology.* Cambridge, England: Cambridge University Press.

Erber, N.P. (1975). Audio-visual perception of speech. *Journal of Speech and Hearing Disorders, 40,* 481–492.

Fant, G. (1960). *Acoustic theory of speech production.* 's-Gravenhage, Netherlands: Mouton and Co.

Finn, K.I. (1986). *An investigation of visible lip information to be used in automated speech recognition.* Unpublished doctoral dissertation, Georgetown University.

Fowler, C.A., Rubin, P., Remez, R.E., & Turvey, M.T. (1978). Implications for speech production of a general theory of action. In B. Butterworth (Ed.), *Language production* (Vol. 1). London: Academic Press.

Fujimura, O. (1961). Bilabial stop and nasal consonants: a motion picture study and its anatomical implications. *Journal of Speech and Hearing Research, 4,* 233–247.

Fujimura, O., Kiritani, S., & Ishida, H. (1973). Computer controlled radiography for observation of movement of articulatory and other human organs. *Computing in Biology and Medicine, 3,* 371–384.

Kiritani, S., Itoh, K., & Fujimura, O. (1975). Tongue-pellet tracking by a computer-aided

X-ray microbeam system. *Journal of the Acoustical Society of America, 57,* 1516–1520.

Kuehn, D.P., Reich, A.R., & Jordan, J.E. (1980). A cineradiographic study of chin marker positioning: Implications for the strain-gauge transduction of jaw movement. *Journal of the Acoustical Society of America, 67,* 1825–1827.

Lewis, J.P., & Parke, F.I. (1987). Automated lip-synch and speech synthesis for character animation. *Proceedings of ACM conference on Computer Human Interaction and Computer Graphics, Toronto* (pp. 143–147). New York: Association of Computing Machinery.

Lindblom, B.E.F., & Soron, H.I. (1965). Analysis of labial movement. *Journal of the Acoustical Society of America, 38,* 935 (abstract).

MacLeod, A., & Summerfield, A.Q. (1987). Quantifying the contribution of vision to speech perception in noise. *British Journal of Audiology, 21,* 131–141.

McCutcheon, M.J., Fletcher, S.G., & Hasegawa, A. (1977). Video-scanning system for measurement of lip and jaw motion. *Journal of the Acoustical Society of America, 61,* 1051–1055.

McGrath, M., Summerfield, A.Q., & Brooke, N.M. (1984). Roles of lips and teeth in lipreading vowels. *Proceedings of the Institute of Acoustics (Autumn Meeting, Windermere), 6*(4), 401–408.

McGrath, M., Summerfield, A.Q., & Brooke, N.M. (1984). Roles of lips and teeth in lipreading vowels. *Proceedings of the Institute of Acoustics (Autumn Meeting, Windermere), 6(4),* 401–408.

McGurk, H., & MacDonald J. (1976). Hearing lips and seeing voices: a new illusion, *Nature, 264,* 746–748.

Montgomery, A.A. (1980). Development of a model for generating synthetic animated lip shapes. *Journal of the Acoustical Society of America, 68,* S58 (abstract).

Montgomery, A.A., & Jackson, P.L. (1983). Physical characteristics of the lips underlying vowel lipreading performance. *Journal of the Acoustical Society of America, 73,* 2134–2144.

Montgomery, A.A., & Soo Hoo, G. (1982). ANIMAT: A set of programs to generate, edit and display sequences of vector-based images. *Behavioural Research Methods and Instrumentation, 14,* 39–40.

Muller, E.M., & Abbs, J.H. (1979). Strain-gauge transduction of lip and jaw motion in the mid-sagittal plane: Refinement of a prototype system. *Journal of the Acoustical Society of America, 65,* 481–486.

Nelson, W.L., Perkell, J.S., & Westbury, R.J. (1984). Mandible movements during increasingly rapid articulations of single syllables: preliminary observations. *Journal of the Acoustical Society of America, 75,* 945–951.

Nishida, S. (1986). Speech recognition enhancement by lip-information. *Proceedings of CHI 86* (pp. 198–204). New York: Association for Computing Machinery.

Parke, F.I. (1975). A model for human faces that allows speech synchronized animation. *Computers and Graphics, 1,* 3–4.

Parke, F.I. (1982). Parametric models for facial animation. *IEEE Computer Graphics and Applications, 2,* 61–68.

Perkell, J.S. (1969). *Physiology of speech production: Results and implications of a quantitative cineradiographic study.* Cambridge, MA: MIT Press.

Petajan, E.D. (1984). Automatic lipreading to enhance speech recognition. *Proceedings of the Global Telecommunications Conference* (pp. 265–272). Atlanta, GA: IEEE Communication Society.

Petajan, E.D., Bischoff, B., Bodoff, D., & Brooke, N.M. (1988). An improved automatic lipreading system to enhance speech recognition. *Proceedings of the CHI 88 Conference* (pp. 19–25). New York: Association for Computing Machinery.

Petajan, E.D., Brooke, N.M., Bischoff, B.J., & Bodoff, D.A. (1988). Experiments in automatic visual speech recognition. *Proceedings of the 7th Symposium of the Federation of Acoustical Societies of Europe* (FASE) (pp. 1163–1170). Edinburgh, Scotland: Institute of Acoustics.

Platt, S.M., & Badler, N.I. (1981). Animating facial expressions. *Computer Graphics, 15,* 245–252.

Sonoda, Y., & Wanishi, S. (1982). New optical method for recording lip and jaw movements. *Journal of the Acoustical Society of America, 72,* 700–704.

Storey, D., & Roberts, M. (1987). Audio-visual speech synthesis with a BBC micro. *Acoustics Bulletin, 12(2),* 7 (abstract).

Summerfield, A.Q. (1987). Some preliminaries to a comprehensive account of audio-visual speech perception. In R. Campbell & B. Dodd (Eds.), *Hearing by eye.* Hillsdale, NJ: Erlbaum.

Summerfield, A.Q., MacLeod, A., McGrath, M., & Brooke, N.M. (1989). Lips, teeth and the benefits of lipreading. In A.W. Young & H.D. Ellis (Eds.), *Handbook of research on face processing.* Amsterdam, Netherlands: North-Holland.

3
The Processing of Face Information in the Primate Temporal Lobe

Edmund T. Rolls*
Oxford University
Department of Experimental Psychology
Oxford, England

INTRODUCTION

In this chapter some of the ways in which information about faces is processed by the brain will be considered. First, empirical studies largely based on recordings of the activity of single neurons in the brain will be described. The recordings in these studies are made mainly in nonhuman primates, because the temporal lobe, in which this processing occurs, is much more developed than in nonprimates, and because the findings are directly relevant to understanding the effects of damage to the temporal lobes in humans. Second, some of the implications of these studies for the computations being performed by neuronal networks in the brain when it processes faces will be considered. Part of the interest of the way in which the brain processes faces is that it provides an existence proof for what can be achieved with this style of computation.

The Stages In Visual Information Processing at which Neurons Respond Selectively to Faces

Visual pathways project by a number of cortico-cortical stages from the primary visual cortex, the striate cortex, until they reach the temporal lobe visual areas (Cowey, 1979; Desimone & Gross, 1979; Seltzer & Pandya, 1978; Maunsell & Newsome, 1987; see Figure 3.1) in which some neurons which respond selec-

---

* The author has worked on some of the experiments described here with P. Azzopardi, G.C. Baylis, M. Hasselmo, L. Hughes, C.M. Leonard, and D.I. Perrett, and their collaboration is sincerely acknowledged. This research was supported by the Medical Research Council.

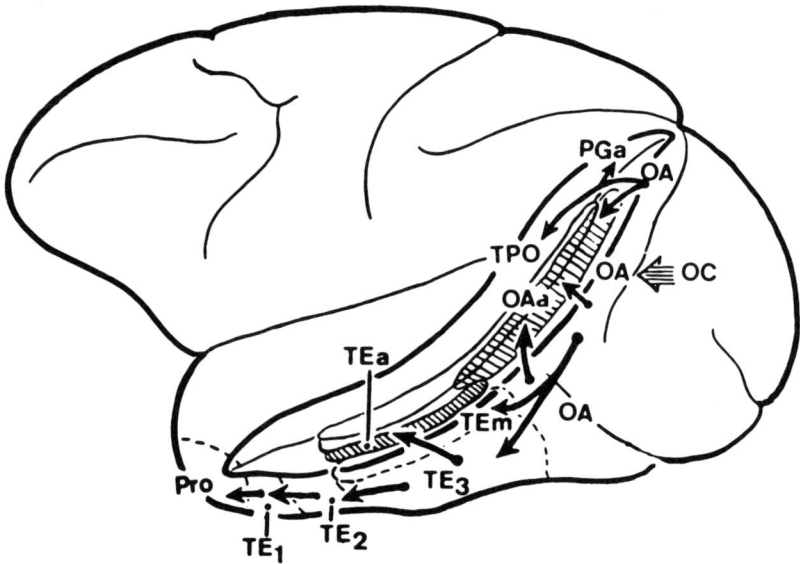

Figure 3.1. Forward projections in the temporal lobe from visual areas of the macaque monkey. OC is the primary (striate) cortex, the TE areas are the inferior temporal visual cortex, and the superior temporal sulcus has been opened. (From Seltzer & Pandya, 1978).

tively to faces are found (Desimone & Gross, 1979; Bruce, Desimone, & Gross, 1981; Desimone, Albright, Gross, & Bruce, 1984; Gross, Desimone, Albright, & Schwartz, 1985; Rolls, 1981a,b, 1984; Perrett, Rolls, & Caan, 1982). The inferior temporal visual cortex, area TE, is divided on the basis of cytoarchitecture, myeloarchitecture, and afferent input into areas TEa, TEm, TE3, TE2, and TE1, and in addition there are a set of different areas in the cortex in the superior temporal sulcus (Seltzer & Pandya, 1978; Baylis, Rolls, & Leonard, 1987) (see Figure 3.2). Of these latter areas, TPO receives inputs from temporal, parietal, and occipital cortex; PGa and IPa from parietal and temporal cortex; and TS and TAa primarily from auditory areas (Seltzer & Pandya, 1978).

In order to investigate the information processing being performed by these parts of the temporal lobe cortex, the activity of single neurons was analyzed in each of these areas in the rhesus macaque monkey (Baylis et al., 1987) during the presentation of simple and complex visual stimuli such as sine wave gratings, three-dimensional objects, and faces; and auditory and somatosensory stimuli. First, it was clear that these areas were influenced differently by inputs from the different modalities, as shown in Figure 3.3. In particular, it was of interest that areas TS and TAa were influenced by visual as well as by auditory inputs; that areas TPO, PGa, and IPa were influenced by somatosensory as well as by visual and auditory inputs; and that areas TEa, TEm, TE3, TE2, and TE1 were influ-

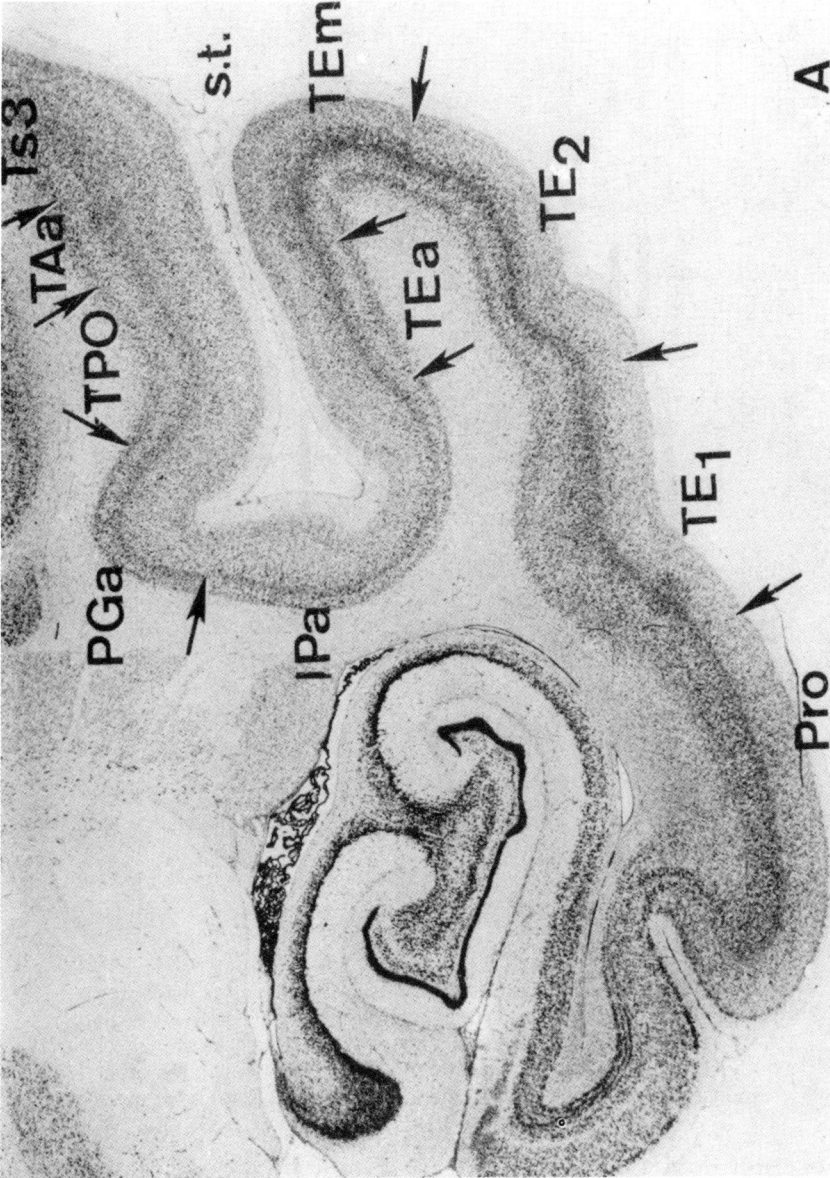

Figure 3.2. Coronal section showing the different architectonic areas in the bordering the anterior part of the superior temporal sulcus of the macaque. The numbers refer to mm posterior to the sphenoid process, which is at the level of the anterior border of the optic chiasm.

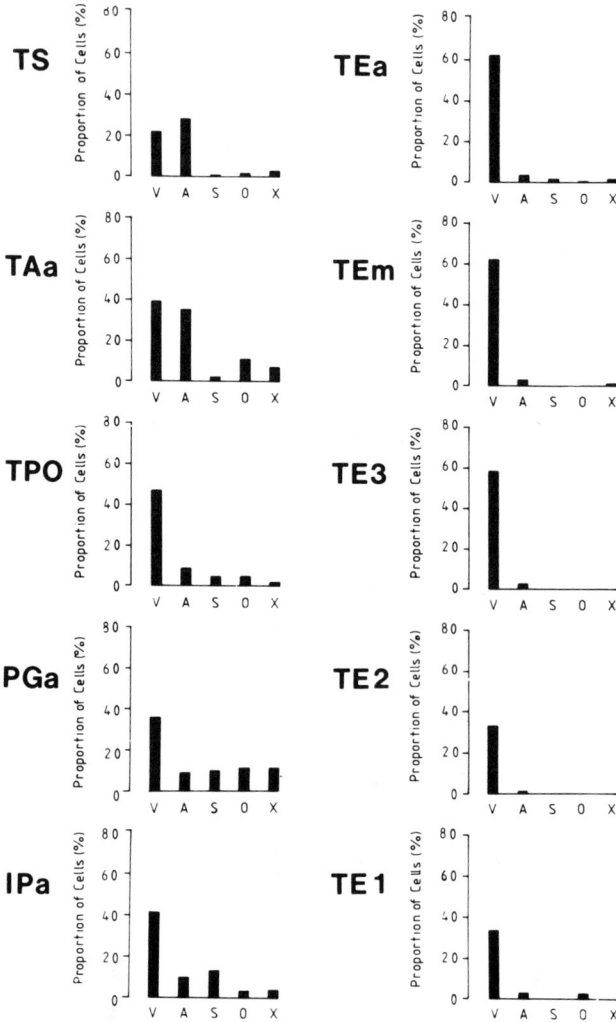

Figure 3.3. Neuronal responses in different areas of cortex within the superior temporal sulcus to stimuli in different modalities. V—Visual; A—Auditory; S—Somatosensory, for example, touch to the leg or body; O—Oral, including mouth movements or touch; X—other responses, including body movement. (From Baylis, Rolls, & Leonard, 1987).

enced primarily by visual inputs (Figure 3.3). Second, it was clear that, within the visual modality, the types of response found were different across these different areas (see Figure 3.4). For example, in areas TS and TAa, most of the visual responses to stationary stimuli were broadly tuned (SB), and some neurons responded only to moving stimuli; in areas TPO, PGa, and IPa, neurons which

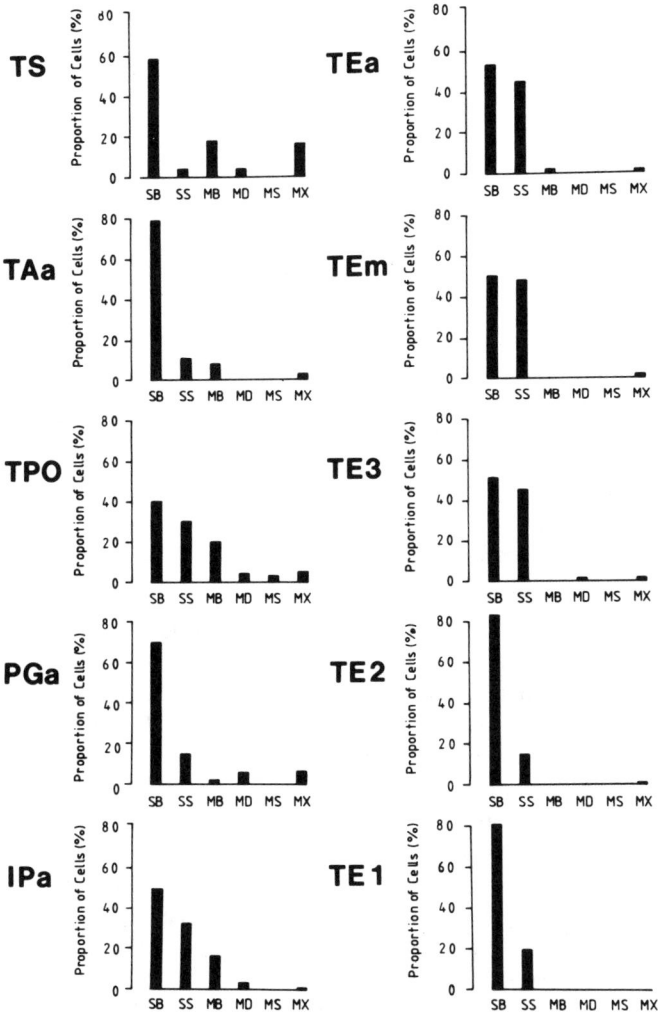

Figure 3.4. Neuronal responses in different areas of cortex within the superior temporal sulcus to different types of visual stimuli. SB—cells responding to stationary stimuli with Broad tuning; SS—cells responding to Stationary stimuli with Selective responses to only some stimuli; MB—cells responding only to Moving visual stimuli, with Broad tuning; MD—cells responding only to Moving visual stimuli, with Direction tuning; MS—cells responding only to Moving visual stimuli, with stimulus selectivity; MX—cells responding only to Moving visual stimuli, with both Directional and Stimulus selectivity. (From Baylis, Rolls, & Leonard, 1987).

responded to stationary stimuli were often selective (SS), and relatively many neurons responded to moving visual stimuli, sometimes with movement, direction, and stimulus selectivity combined (MX); and in areas TEa–TE1, relatively many neurons responded to stationary stimuli with stimulus selectivity (SS), and few neurons responded to moving visual stimuli (see Figure 3.4). Third, it was clear that the neurons which responded with stimulus selectivity were localized with different types of selectivity predominant in the different architectonic areas (see Figure 3.5). In particular, neurons which responded primarily to faces were found predominantly in areas TPO, TEa, and TEm, whereas neurons which responded to simple visual stimuli such as sine wave gratings (G) were relatively more common in areas TE3 and IPa, and neurons which responded only to complex visual stimuli were relatively common in areas TE1, TE3, and TEm (see Figure 3.5). These and the other differences between areas evident in Figures 3.3–3.5 thus make it clear that different types of information processing are occurring in these different areas in the temporal lobe cortex, and that therefore functional localization occurs at least down to this level.

The Selectivity of One Population of Neurons for Faces

The findings just summarized show that neurons which respond selectively to faces are not simply the most complex cells found throughout the inferior temporal visual cortex, but instead appear to be a specialized population in that they are found particularly in cytoarchitectonic areas TPO, TEa, and TEm (Baylis et al., 1987). The degree of their selectivity for faces is considered next. The responses of these neurons to faces are selective in that they are 2–10 times as large to faces as to gratings, simple geometrical stimuli, or complex 3-D objects (Perrett et al., 1982; Rolls, 1984; Baylis, Rolls, & Leonard, 1985, 1987). Two particular criteria are used to select neurons with responses which occur primarily to faces for analysis. First, their responses to one or more faces must be at least twice as large as to any other of the wide range of visual stimuli tested. (In fact, the majority of the neurons in the cortex in the superior temporal sulcus classified as showing responses selective for faces responded much more specifically than this. For half these neurons, their response to the most effective face was more than five times as large as to the most effective nonface stimulus, and for 25 percent of these neurons, the ratio was greater than 10:1. (The degree of selectivity shown by different neurons studied is illustrated in Figure 3.6.) Second, an analysis of variance performed over the set of face and nonface stimuli must show a significant effect of stimulus type, and subsequent multiple t, Tukey, and Newman-Keuls' analysis (see Bruning & Kintz, 1977) must show that the response to the optimal face stimulus was significantly greater ($p < 0.05$) than the response to the optimal nonface stimulus. (In fact, the difference was signifi-

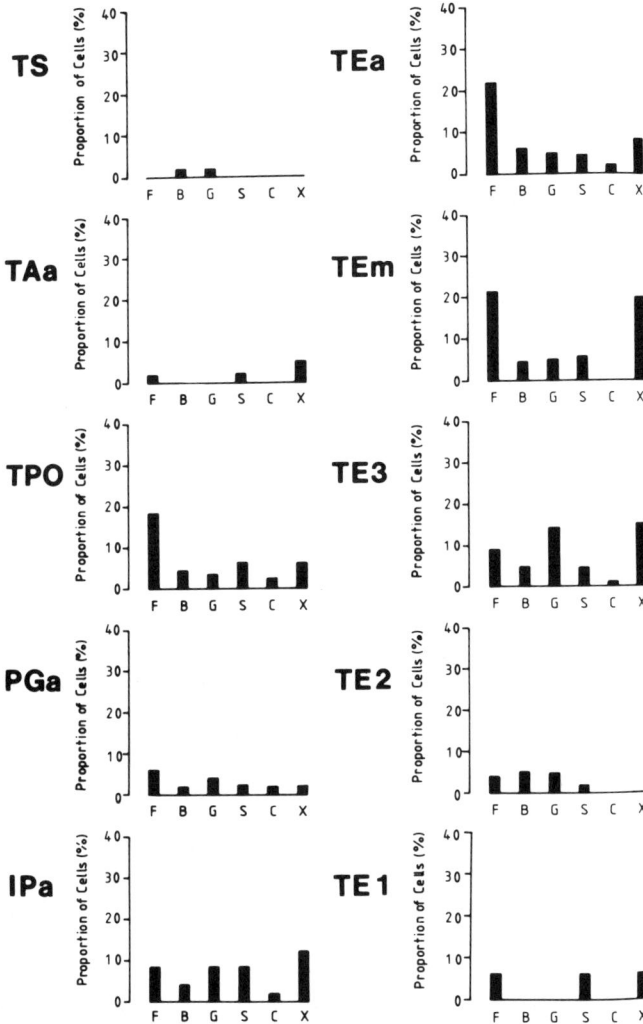

Figure 3.5. Neuronal responses in different areas of cortex within the superior temporal sulcus to different patterns of visual stimuli. F—respond selectively to faces; G—major response and tuned to Boundary curvature descriptors; G—major response and tuned to Sine wave grating; S—Shape selectivity; C—Color selectivity; X—response only to complex stimulus. (From Baylis, Rolls, & Leonard, 1987).

cant at the 0.01 level for the great majority of the neurons studied.) The responses to faces are excitatory, sustained and are time-locked to the stimulus presentation with a latency of between 80 and 160 ms. The cells are typically

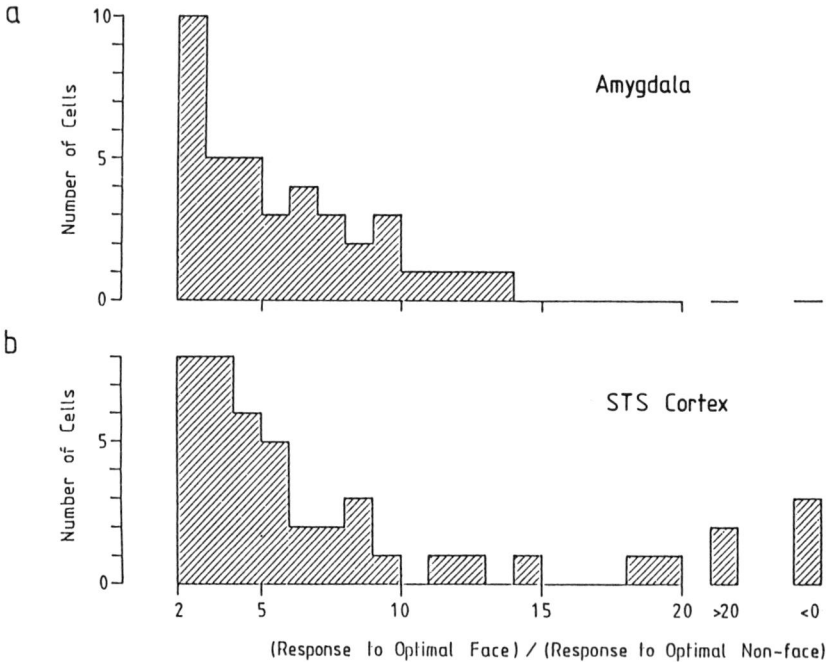

Figure 3.6. The degree of selectivity of some neurons recorded in the cortex in the superior temporal sulcus (below) and amygdala (above) is indicated by the ratio of the response of each neuron to its best face stimulus, to its response to its best nonface stimulus. It is clear that these neurons do respond much better to faces than to nonfaces, but that the selectivity is not absolute. Thus the neurons are like well-tuned but not perfect filters.

unresponsive to auditory or tactile stimuli and to the sight of arousing or aversive stimuli. The magnitude of the responses of the cells is relatively constant despite transformations such as rotation so that the face is inverted or horizontal, and alterations of color, size, distance, and contrast (Perrett et al., 1982; Rolls & Baylis, 1986). Thus these neurons at this high level of visual information processing have some properties of perceptual invariance which relate them closely to perception (Rolls & Baylis, 1986). Masking out or presenting parts of the face (e.g., eyes, mouth, or hair) in isolation reveal that different cells respond to different features or subsets of features. For some cells, responses to the normal organization of cut-out or line-drawn facial features are significantly larger than to images in which the same facial features are jumbled (Perrett et al., 1982). These findings indicate that explanations in terms of arousal, emotional, or motor reactions, and simple visual feature sensitivity or receptive fields, are insufficient to account for the selective responses to faces and face features observed in this population of neurons in the cortex in the superior temporal sulcus (Perrett et al., 1982; Baylis et al., 1985; Rolls & Baylis, 1986). Observations consistent with these findings have been published by Desimone et al. (1984), who described a similar population of neurons located primarily in the cortex in the superior temporal sulcus which responded to faces but not to simpler stimuli such as edges and bars, or to complex nonface stimuli (see also Gross et al., 1985).

Ensemble Encoding of Face Identity

An important question for understanding brain function is whether a particular object (or face) is represented in the brain by the firing of one or a few gnostic or "grandmother" cells, or whether, instead, the firing of a group or ensemble of cells each with somewhat different responsiveness provides the representation (Barlow, 1972). We have investigated whether the face-selective neurons encode information which could be used to distinguish between faces, and if so, whether the neurons are gnostic, or whether, instead, ensemble encoding is used. We have found that, in many cases (77% of one sample), these neurons are sensitive to differences between faces (as shown by an analysis of variance) (Baylis et al., 1985). However, each neuron does not respond only to one face. Instead, each neuron has a different pattern of responses to a set of faces, as illustrated in Figure 3.7.

To quantify the extent to which a neuron responded differently to different faces, the difference between the response to the most effective face stimulus and the least effective face stimulus (both averaged over all 4–10 presentations of that face) was calculated and presented as a ratio to the standard deviation of the responses. This measure thus represents the number of standard deviations which separate the two neuronal responses and is intended to be analogous to detectability, d', in signal detection theory (see Green & Swets, 1966; Egan, 1975; Rolls, Thorpe, Boytim, Szabo, & Perrett, 1984). This measure has also been chosen so that it may be compared with d' measures of the discriminability of faces to human observers. The values of d' for the neurons tested in this way are shown in Figure 3.8a (see Baylis et al., 1985). It is clear that a range of degrees of selectivity for different faces was found, with some neurons responding quite well even to the least effective face (values of d' less than 1.0), and with some neurons responding very differently to the different faces in the set (values of d' greater than 1.0). It is clear that, over the population, a great deal of information which could be used to discriminate between the faces of different individuals is present in the responses of these neurons.

To quantify how finely these neurons were tuned to the faces of particular individuals, a measure derived from information theory, the breadth of tuning metric developed by Smith and Travers (1979), was calculated. This is a coefficient of entropy (H) for each cell which ranges from 0.0, representing total

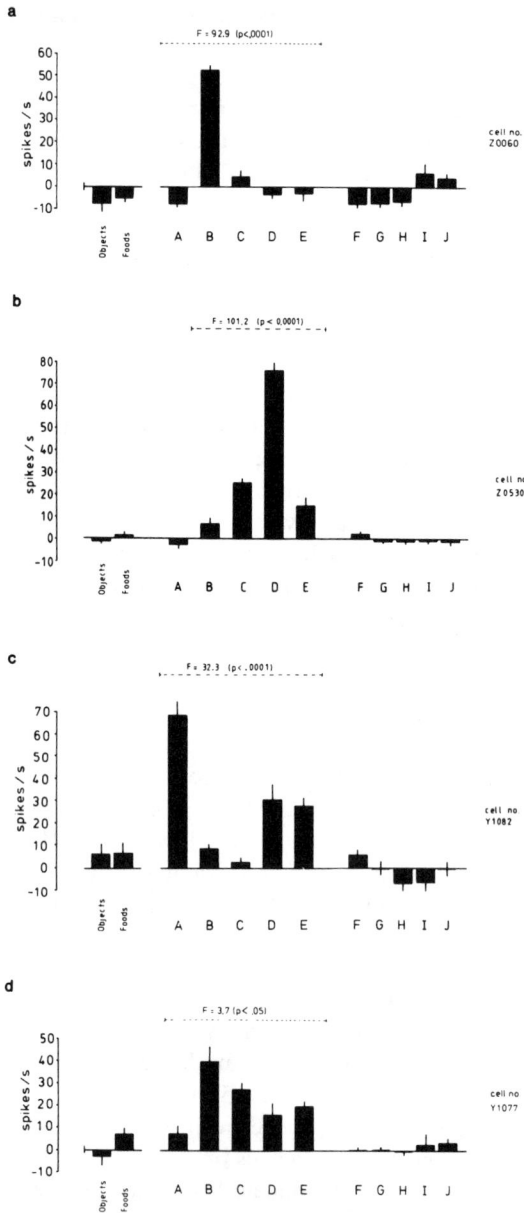

Figure 3.7. The responses of four cells (a–d) in the cortex in the superior temporal sulcus to a variety of face (A–E) and nonface (F–J) stimuli. The bar represents the mean firing rate response above the spontaneous baseline firing rate with the standard error calculated over 4–10 presentations. The F ratio for the analysis of variance calculated over the face sets indicates that the units shown range from very selective (Z0060) to relatively nonselective (Y1077). (From Baylis, Rolls, & Leonard, 1985).

specificity to one stimulus, to 1.0, which indicates an equal response to the different stimuli.[1]

The breadth of tuning of the population of neurons analyzed in this way is shown in Figure 3.8b. It is clear that the neurons did not respond only to the face of one individual, but that, instead, typically each neuron responded to a number of faces in the stimulus set (which included five different faces). This is also shown by the generalization index in Figure 3.8c. The generalization index is the proportion of the faces in the set of faces to which the neuron had a response greater than half that to the most effective face stimulus. In another investigation it was again found that neurons which respond differently to the faces of different individuals (and were not therefore responding just to the presence of eyes or on the basis of emotional expression) do so using ensemble encoding (Hasselmo, Rolls, & Baylis, 1989). It may be noted that there are at present no quantitative measures of the breadth of tuning of neurons available to support the grand-mother cell method of encoding information in these populations of neurons (Perrett, Mistlin, & Chitty, 1987).

The evidence described above shows that the responses of each of these neurons in the cortex in the superior temporal sulcus does not code uniquely for the face of a particular individual. Instead, across a population of such cells information is conveyed that would be useful in making different behavioral responses to different faces. Thus information which specifies an individual face is present across an ensemble of such cells. In that each neuron does not respond to only one face, and in that a particular face can activate many neurons, these are not "grandmother" cells (Barlow, 1972). However, in that their responses are relatively specialized both for the class "face" and within this class, they could contribute to relatively economic coding of information over relatively few cells (Barlow, 1972). It may be noted that, even if individual neurons in this population are not tuned to respond completely specifically to only face stimuli, it is nevertheless the case that the output of such an ensemble of neurons would be useful for distinguishing between different faces and would provide a unique identification of that individual. Expressed quantitatively, n neurons could be considered as an n-dimensional vector capable of representing a unique point in an n-dimensional space if the neurons have independent responses, or as representing a unique point in a space of somewhat less than n dimensions if the neuronal responses are somewhat dependent.

These findings thus lead to the hypothesis that these neurons are filters, the output of which taken as an ensemble could be used for recognition of different individuals, and in emotional responses made to different individuals. Their

---

[1] $H = -k \sum_{i=1}^{n} p_i \log p_i$

where H = breadth of responsiveness, K = scaling constant (set so that H = 1.0 when the neuron responds equally well to all stimuli in the set of size n), and p_i = the response to stimulus i expressed as a proportion of the total response to all the stimuli in the set.

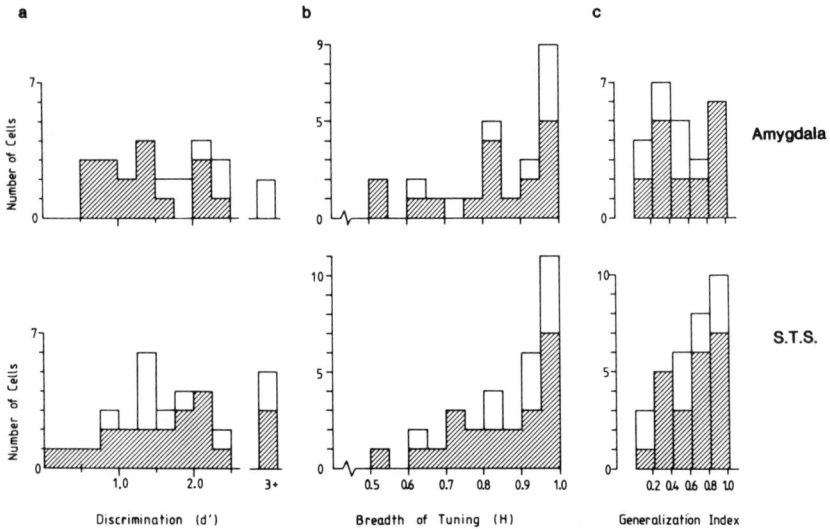

Figure 3.8. (a): The discriminability indices (d′); (b): the breadths of tuning; (c): the generalization indices calculated across the standard set of faces for cells recorded in the cortex in the superior temporal sulcus. (From Baylis, Rolls, & Leonard, 1985).

different responses, not only to different faces, but also to different parts of faces (Perrett et al., 1982) and to different parts of the spatial frequency spectrum present in faces (Rolls, Baylis, & Leonard, 1985; Rolls, Baylis, & Hasselmo, 1987) provide further evidence for understanding them as filters.

It may also be noted that it is unlikely that there are further processing areas beyond those described where ensemble coding changes into grandmother cell encoding, in that, anatomically, there does not appear to be a whole further set of visual processing areas present in the brain; and from the temporal lobe visual areas such as those described, outputs are taken to limbic and related regions such as the amygdala and via the entorhinal cortex the hippocampus. Indeed, tracing this pathway onwards, Leonard, Rolls, Wilson, and Baylis (1985) have found a population of neurons with face-selective responses in the amygdala, and in the majority of these neurons, different responses occur to different faces, with ensemble not unique coding still being present. After interfacing with limbic circuits in this way, there is evidence that there are further links which may be important in the behavioral output via the connections of the amygdala (Rolls, 1985) to the ventral striatum (which includes the nucleus accumbens) (see Rolls & Williams, 1987), for in the ventral striatum a small number of neurons is found which also respond to faces (see Rolls & Williams, 1987).

The Advantages of Ensemble Encoding

Given this empirical evidence for ensemble encoding, we can ask what the advantages are of this type of coding in which the representation is distributed over many neurons. The advantages can be understood by considering information representation and storage in networks of neurons, such as that shown in Figure 3.9. This particular network achieves a form of association memory, such that, when a conditioned stimulus pattern (such as the sight of food) is placed on the input axons, and simultaneously an unconditioned stimulus (such as the taste of the food) is placed on the forcing stimulus lines to make the output neurons fire, the synapses between the active input axons and the simultaneously active vertical dendrites increase in strength. This learning rule is usually called Hebbian. After this learning phase, subsequent presentation of only the conditioned stimulus along the axons comes to make the output neurons fire in the same pattern as they originally fired to the unconditioned stimulus. This occurs because of the particular synapses which were modified during the learning phase according to the Hebbian rule. This type of simple network shows a number of interesting properties, which are sometimes called *emergent*. One of these is *completion*. If only part of the conditioned stimulus is given (as may often happen to real organisms), then the network produces as output, not part of the unconditioned stimulus, but all of it. This is completion. A second property is *generalization,* which occurs in that, if a stimulus similar but not necessarily

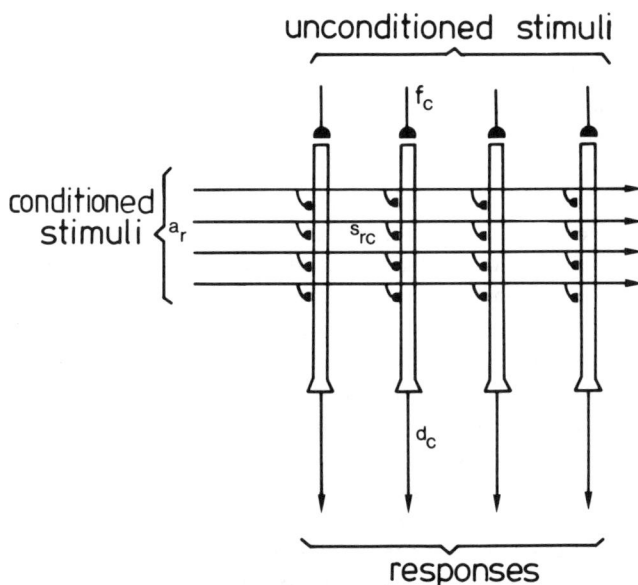

Figure 3.9. Neurons connected to form an association matrix memory. The vertical rectangles represent the dendrites of the neurons which respond unconditionally to application of the conditioned or forcing stimuli (f) to produce the responses (d). The conditioned stimuli are applied to the horizontally running axons (a), each one of which synapses with each dendrite it passes with a modifiable strength (s) (see text).

identical to the original conditioned stimulus is applied, the response recalled is very similar to that recalled to the original conditioned stimulus. This is generalization. A third property is *graceful degradation,* in that, if part of the network is damaged, then the response recalled will still approximate the correct response. That is, the recall degrades only gradually and gracefully (as contrasted with catastrophically) when the network (or brain) is damaged. These properties are described more fully elsewhere (Kohonen, 1988; Rumelhart & McClelland, 1986; Rolls, 1987a,b). The important point here is that these properties of neuronal networks are only found if the inputs are ensemble encoded, that is, if several input neurons are used to represent any one stimulus or event in the environment. It is suggested that this is why ensemble encoding is used even far on in the visual system, and is why "grandmother cell" encoding is not used. On the other hand, each event must not be represented over a very large population of neurons which overlaps almost completely with the population activated by a different event, for, if this were the case, the matrix memory would display great interference and would be a very inefficient memory storage system. Given that neurons have positive firing rates which appear in many parts of the brain against a very low or zero level of spontaneous firing, the only way in which the relatively orthogonal representations required can be formed is by making the number of neurons active for any one input stimulus relatively low (see, e.g., Jordan, 1986). With sparse representations a large number of different patterns can be stored in the memory (see Rolls, 1987a,b, 1989a,b; Rolls & Treves, 1990).

Thus, the answer we have reached for how information is represented in the nervous system is that neurons become sharply tuned far on in sensory processing but yet have overlap in their tuning, so that the activity of an ensemble of neurons is needed to individuate an item in the environment. The type of tuning found is thus seen as a delicate compromise between very fine tuning, which has the advantage of low interference in neuronal network operations but the disadvantage of losing the emergent properties of storage in neuronal networks, and broad tuning, which has the advantage of allowing the emergent properties of neuronal networks to be realized but of leading to interference between the different memories stored in the network (see Rolls, 1987a,b, 1989a,b). Moreover, it is appropriate that the representations in the inferior temporal cortex are in distributed form ready for association neuronal networks, for the outputs of the temporal lobe visual areas project to structures such as the amygdala and orbitofrontal cortex which are involved in cross-modal association memory functions, and to structures such as the hippocampus which contain neuronal networks which may implement episodic memory by using an autoassociation neuronal network (Rolls, 1987a,b, 1989a,b,c, 1990).

The Development of Specificity of the Neuronal Responses

Given that the responses of these neurons to faces do show considerable stimulus selectivity, we may ask how this specificity is achieved by the nervous system.

We may note that the computation of a representation with relatively finely tuned neurons appears to be an important computation to be performed by any sensory system in the brain. For example, in the taste system, it is found that the breadth of tuning of single neurons to the basic tastes sweet, salt, bitter, and sour decreases as populations of neurons are traced from the first central synapse in the nucleus of the solitary tract, through the primary taste cortex in the frontal operculum and insula, to the secondary taste cortex in the caudolateral orbitofrontal cortex (Rolls, 1986c, 1989d).

Given the fundamental importance of this type of computation, which results in relatively finely tuned neurons which, across ensembles but not individually, specify objects in the environment, we have investigated whether experience plays a role in determining the selectivity of single neurons which respond to faces. The hypothesis being tested was that visual experience might guide the formation of the responsiveness of neurons so that they provide an economical and ensemble-encoded representation of items actually present in the environment. To test this, we investigated whether the responses of the face-selective neurons were at all altered by the presentation of new faces which the monkey had never seen before. It might be, for example, that the population would make small adjustments in the responsiveness of its individual neurons, so that neurons would be provided with filter properties which would enable the population as a whole to discriminate between the faces actually seen. Indeed, in systems which use a distributed representation, it is implied that many neurons may alter their responsiveness a little when a new stimulus is learned. This problem has been scarcely addressed in the context of the real nervous system. We thus asked whether, when a set of totally novel faces was introduced, the responses of these neurons were fixed and stable from the first presentation, or, instead, whether there was some adjustment of responsiveness over repeated presentations of the new faces. First, it was shown for each neuron tested that its responses were stable over 5–15 iterations of a set of familiar faces. Then a set of new faces was shown in random order, and the set was repeated with a new random order over many iterations. It was found that some of the neurons studied in this way altered the relative degree to which they responded to the different members of the set of novel faces over the first few (1–2) presentations of the set (Rolls, Baylis, Hasselmo, & Nalwa, 1989). Thus, there is now some evidence from these experiments that the response properties of neurons in the temporal lobe visual cortex are modified by experience, and that the modification is such that, when novel faces are shown, the relative responses of individual neurons to the new faces alter. It is hypothesized that alteration of the tuning of individual neurons in this way results in a good discrimination over the population as a whole of the faces known to the monkey.

In searching for models to help understand better the representation of information across ensembles of neurons, and how optimal discrimination of the inputs actually presented is developed by the population, we have found neuronal net models with competition to be helpful. An example of one type of competitive neuronal net which achieves useful categorization is described next (for further details see Rolls, 1989a,b,e).

Consider a neuronal network in which the strengths of the synapses between horizontal axons and the vertical dendrites are initially random. (Alternative designs are to make the probability that a horizontal axon makes a synapse with every (vertical) neuron it passes not near 1, as in Figure 3.10, but much lower, perhaps in the range 0.1–0.001; or, alternatively, to allow every axon to make a random number of contacts with a dendrite.) The result is that different input patterns on the horizontal axons will tend to activate different neurons. The tendency for each pattern to select or activate different neurons can then be enhanced by (a) providing mutual inhibition between the output neurons to enhance the selectivity with which each neuron responds, and then (b) increasing the synaptic weights at those synapses where there is both pre- and postsynaptic activity (cf. long-term potentiation, Bliss & Lomo, 1973; Andersen, 1987; McNaughton, 1984; Levy, 1985; Levy & Desmond, 1985; Kelso, Ganong, & Brown, 1986; Wigstrom, Gustaffson, Huang, & Abraham, 1986). The mutual inhibition is one form of competition and helps prevent too many neurons becoming allocated to any one stimulus. An example of the operation of such a competitive network to produce categorization is shown in Figure 3.11 (for fuller details, see Rolls, 1990). Such a categorization process effectively selects different neurons to respond to different combinations of active input horizontal lines. It should be noted that this categorization finds natural clusters in the input events; orthogonalizes the input events, in that overlap in input events can become coded onto output neurons with less overlap, and in that many active input lines may be coded onto few active output lines; and does not allocate neurons to events which never occur (cf. Marr, 1970, 1971; Grossberg, 1987; Rumelhart &

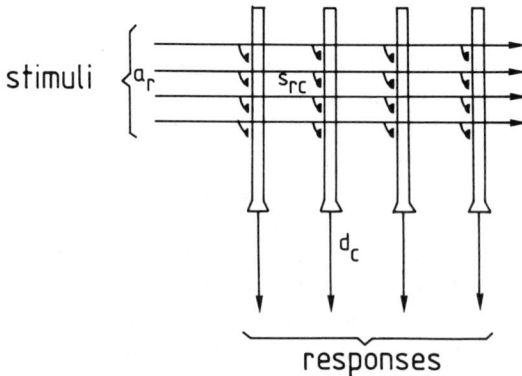

Figure 3.10. Neurons connected in a competitive neuronal net. There is one input received by the axons (a), which are connected to the dendrites (d) to the output neurons by modifiable synapses (s) (see text).

Zipser, 1985; Rolls, 1989a,b,e). The small modifications in the responses of neurons over the first few presentations of novel face stimuli are consistent with a form of active tuning of the responses of individual neurons being performed in a population of neurons interacting in this competitive way to achieve good classification.

In a recent development of our simulations of competitive neuronal networks we have incorporated some of the nonlinearity in the synaptic modification rule which results from the operation of the NMDA receptors, and have found it beneficial to the operation of the competitive networks. The learning rule used is

$$\delta s_{rc} = k.m_c(a_r - s_{rc})$$

where k is a constant, δs_{rc} is the change of synaptic weight, a_r is the firing rate of the r'th axon, and m_c is a nonlinear function of the activation of the dendrite d_c which mimics the operation of the NMDA receptors in learning as follows (Cotman, Monaghan, & Ganong, 1988). Long-term potentiation occurs when the postsynaptic membrane becomes so strongly activated and thus depolarized that the NMDA receptors, which are voltage-sensitive, are activated. Activation of the NMDA receptors then allows Ca^{++} to enter the cell, which is necessary for the long-term change in synaptic efficacy (see Cotman et al., 1988). There is thus a strong nonlinearity in synaptic modification, which only occurs for neurons which are strongly activated. We have simulated this by, for example, setting m_c to be proportional to the square of d_c, and rescaling to the previous maximum firing rate. The decay term (cf. Levy & Desmond, 1985) results in a decay of weakly active synapses onto strongly activated output neurons. We (see Rolls, 1989e) have found in simulations that competitive networks categorize well with this rule, and that the rule can maintain the lengths of the synaptic weight vectors to be very similar without any need for any other explicit normalization of the synaptic strengths. Because of the nonlinearity in the learning rule, the competition between the output neurons need not be extremely strong, with categorization occurring well when, for example, the output firing rates are raised to only a power of 2 before rescaling the output firing rates to a given maximum and minimum. This highlights a major advantage of the NMDA learning rule for competitive neuronal networks, for the nonlinearity in the learning rule means that a more physiologically plausible degree of competition between the output neurons is sufficient for good categorization to occur. The rule can be reexpressed as

$$\delta \mathbf{s}_c = k.m_c(\mathbf{a} - \mathbf{s}_c)$$

where $\delta \mathbf{s}_c$ is the change in the synaptic weight vector on neuron c with activation of its NMDA receptors m_c, and $\mathbf{a}$ is the input pattern vector on the axons, which makes it clear that this rule tends to move the dendritic weight vector in such a

INPUT: | CYCLE: 202

COMPETITION MATRIX

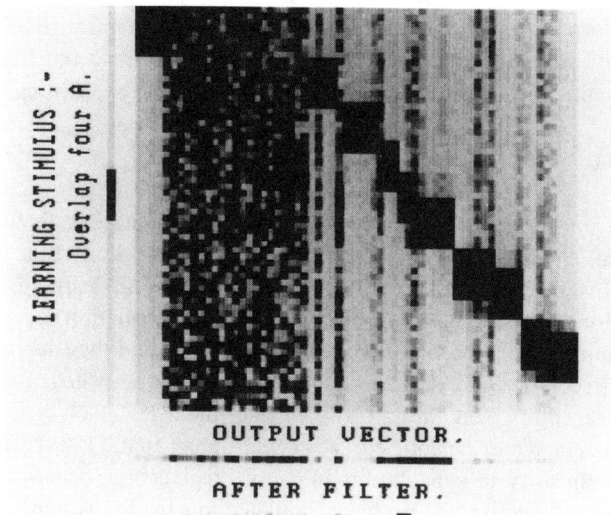

LEARNING STIMULUS :-
Overlap four A.

OUTPUT VECTOR.

AFTER FILTER.

INPUT:]   CYCLE: 202

COMPETITION MATRIX

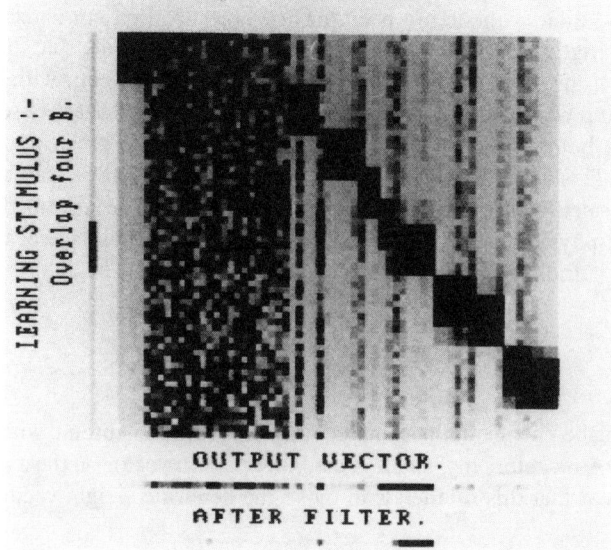

LEARNING STIMULUS :-
Overlap four B.

OUTPUT VECTOR.

AFTER FILTER.

way that it rapidly becomes aligned with the input pattern vector which strongly activates that neuron.

In developing these models further, we have incorporated in multilayer systems backprojections of the type found between adjacent cortical areas which are a major feature of the connectivity of the neocortex (see Rolls, 1989a,b). The guidance these provide helps the competitive networks to learn solutions which can be guided by what can only be detected at the next layer in the hierarchy, and may be useful in building some of the representations which help in solving some of the computational problems which must be solved by the visual system, such as translation and size invariance. The evidence that the representation of information about face identity in the neurons described here does incorporate considerable invariance is described next.

A Neuronal Representation of Faces Showing Invariance

One of the major problems which must be solved by a visual system is the building of a representation of visual information which allows recognition to occur relatively independently of size, contrast, spatial frequency, angle of view, and so on. To investigate whether these neurons in the temporal lobe visual cortex are at a stage of processing where such invariance is being represented in

---

←

Figure 3.11. Simulation of learning in a competitive matrix memory. The architecture is as shown in Figure 10, except that there are 64 horizontal axons and 64 vertical dendrites which form the row and columns respectively of the 64 × 64 matrix of synapses. The strength of each synapse, which was initially random, is indicated by the darkness of each pixel. The activity of each of the 64 input axons is represented in the 64 element vector at the left of the diagram by the darkness of each pixel. The output firing of the vertical neurons is represented in the same way by the output vectors at the bottom of the diagram. The upper output vector is the result of multiplying the input stimulus through the matrix of synaptic weights. The vector resulting from the application of competition between the output neurons (which produces contrast enhancement between the elements of neurons of the vector) is shown below by the vector labeled "after filter." The state of the matrix is shown after 202 cycles, in each of which stimuli with eight of 64 active axons were presented, and the matrix allowed to learn as described in the text. The stimuli were presented in random sequence, and consisted of a set of vectors which overlapped in 0, 1, 2, 3, 4, 5, or 6 positions with the next vector in the set. The columns of the matrix were sorted after the learning to bring similar columns together, so that the types of neuron formed, and the pattern of synapses formed on their dendrites, can be seen easily. The dendrites with random patterns of synapses have not been allocated to any of the input stimuli. It is shown that application of one of the input stimuli (overlap four A) or vectors which overlapped in 4 of 8 positions with another stimulus (overlap four B) produced one pattern of firing of the output neurons, and that application of input stimulus overlap four B produced a different pattern of firing of the output neurons. Thus the stimuli were correctly categorized by the matrix as being different.

the responses of neurons, the effect of such transforms of the visual image on the responses of the neurons has been investigated.

To investigate whether the responses of these neurons show some of the perceptual properties of face recognition such as tolerance to changes in the size and contrast of the face, the effects of alteration of the size and contrast of an effective face stimulus on the responses of these neurons were analyzed quantitatively in macaque monkeys (Rolls & Baylis, 1986). First, it was shown that the majority of these neurons had responses which were relatively invariant with respect to the size of the stimulus. The median size change tolerated with a response of greater than half the maximal response was 12 times. Second, it was found that, for a few of these neurons, the size of the face did affect the neuronal response. In most of these neurons, it was found that, when the size of the image and its distance were altered, the neuronal response was related to the retinal angle subtended by the image. But for four neurons the absolute size of the image determined the magnitude of the neuronal response, independently of the distance of the image. Thus these four neurons showed size constancy. It is suggested that these neurons would be useful as part of a face recognition system, because only objects in a certain absolute size range should normally be classified as faces. Third, the responses of the neurons were relatively invariant with respect to the contrast of the face. The mean contrast at which the neurons still responded with more than half the maximal response was 0.26. Fourth, the responses of the neurons were relatively invariant with respect to the sign of the contrast of the face, that is, the neurons responded to negative as well as to positive images of faces. Fifth, the neurons typically responded to a face when the information in it had been reduced from 3-D to a 2-D representation in gray on a monitor, with a response which was on average 0.5 that to a real face. These results show that the responses of these neurons have some of the invariant properties with respect to size and contrast alteration shown by face perception, and show that their processing is at a level which would be useful in face recognition (Rolls & Baylis, 1986). The results also show that the responses of these neurons are not simply to local contour information.

To investigate whether neurons in the inferior temporal visual cortex and cortex in the anterior part of the superior temporal sulcus operate with translation invariance when these cortical regions are operating normally in the awake behaving primate, their responses were measured during a visual fixation (blink) task in which stimuli could be placed in different parts of the receptive field (Azzopardi & Rolls, 1989). It was found that, in many cases, the neurons responded (with a greater than half-maximal response) even when the monkey fixated 2–5 degrees beyond the edge of a face which subtended 3–20 degrees at the retina. Moreover, the stimulus selectivity between faces was maintained this far eccentric within the receptive field. These results held even across the visual midline. It is concluded that at least some of these neurons in the temporal lobe visual areas do have considerable translation invariance so that this is a computation which must be performed in the visual system.

Another transform over which recognition is relatively invariant is spatial frequency. For example, a face can be identified when it is a line drawing (in which only high spatial frequencies are present), and when it is blurred (when it contains only low spatial frequencies). It has been shown that, if the face images to which these neurons respond are low-pass filtered in the spatial frequency domain (so that they are blurred), many of the neurons still respond when the images only contain frequencies up to 8 cycles per face. Similarly, the neurons still respond to high-pass filtered images (with only high spatial frequency edge information) when only frequencies down to 8 cycles per face are included (Rolls et al., 1985). Now, it is known that face recognition can still proceed with either faces low-pass filtered up to 8 cycles per face, or high-pass filtered down to 8 cycles per face. Thus, the responses of these neurons closely parallel human face recognition under these conditions. Further analysis of these neurons with narrow (octave) bandpass spatial frequency filtered face stimuli shows that the responses of these neurons to an unfiltered face cannot be predicted from a linear combination of their responses to the narrow band stimuli, so that these neurons do not have linear properties (Rolls et al., 1987). The lack of linearity of these neurons, and their responsiveness to a wide range of spatial frequencies, indicate that, in at least this part of the primate visual system, recognition does not occur using Fourier analysis of the spatial frequency components of images (Campbell, 1983; Pinker, 1984).

An Object-Centered Representation of Visual Information

The finding that the responses of these neurons show considerable invariance over changes in the size and contrast of face stimuli is evidence that the encoding is not in retinal coordinates. Alternatives are that it is in viewer-centered or object-centered coordinates.

Some neurons with responses selective for faces only respond if the face is moving (Perrett et al., 1985a). It is noted that encoding of gestures is likely to be important in the social behavior of primates. We took advantage of the fact that these neurons respond to moving faces to investigate whether the encoding of faces by these neurons is in viewer-centered or object-centered coordinates (Hasselmo, Rolls, & Baylis, 1986a; Hasselmo, Rolls, Baylis, & Nalwa, 1989). For 10 neurons it has been shown that the neuron responds to particular movements which can only be described in object-centered coordinates. For example, four neurons responded vigorously to a head undergoing ventral flexion, irrespective of whether the view was full face, of either profile, or even of the back of the head. These different views could only be specified as equivalent in object-centered coordinates. Further, for all of the neurons that were tested in this way, the movement specificity was maintained across inversion, responding, for example, to ventral flexion of the head irrespective of whether the head was upright or inverted. In this procedure, retinally encoded or viewer-centered movement vectors are reversed, but the object-centered description remains the same. It was of interest that the neurons tested generalized across different heads performing the same movements.

Further evidence supporting the hypothesis that some of the neurons in this region use object-centered descriptions is that their selectivity between the faces of different individuals is maintained across anisomorphic transforms of the stimulus. For example, some neurons reliably responded differently to the faces of two different individuals independently of viewing angle. However, in most cases (16/18 neurons), although the identity of the face was reflected in the neuronal response, the response was not perfectly view-independent, and viewing angle also influenced the response. It is possible that these latter neurons represent an intermediate stage in the computation of object-centered descriptions (Hasselmo et al., 1989).

Also consistent with object-centered encoding is the finding of neurons which respond to images of faces of a given absolute size, irrespective of the retinal image size. These neurons, described above, thus show size constancy (Rolls & Baylis, 1986).

Different Neurons are Specialized for Recognition and for Face Expression Decoding

To investigate whether there are neurons in the cortex in the anterior part of the superior temporal sulcus of the macaque monkey which could provide information about facial expression (Rolls, 1981b, 1984, 1986a,b), neurons were tested with face stimuli which included examples of the same individual monkey with different facial expressions (Hasselmo, Rolls, & Baylis, 1986b, 1989). The responses of 45 neurons with responses selective for faces were measured to a set of three individual monkey faces with three expressions for each individual, as well as to human expressions presented through a wide aperture shutter. Of these neurons, 15 neurons showed response differences to different identities independently of expression, and nine neurons showed responses which depended on expression but were independent of identity, as measured by a two-way ANOVA. The neurons responsive to expression were found primarily in the cortex in the superior temporal sulcus, while the neurons responsive to identity were found in the inferior temporal gyrus. These results show that there are some neurons in this region the responses of which could be useful in providing information about facial expression, of potential use in social interactions in groups (Rolls, 1981b, 1984, 1986a,b). Damage to this population may contribute to the deficits in social and emotional behavior which are part of the Kluver-Bucy syndrome produced by temporal lobe damage in monkeys (Rolls, 1981b, 1984, 1986a,b; Leonard et al., 1985).

Synthesis

The evidence described above is consistent with the view that a number of important high-level operations in visual information processing take place in the temporal lobe visual areas. One may be the building of object-centered descriptions of visual information, for in addition to neurons which represent information in viewer-centered coordinates, such as neurons which respond differently to a profile and to a full face, there are neurons which respond in object-centered coordinates and neurons which are intermediate between these representations. It is our hypothesis that this part of the brain builds object-centered descriptors from viewer-centered descriptors, using a mechanism which includes competition and backprojections in the way summarized above and elsewhere (Rolls, 1989a,b). It is also our hypothesis that similar mechanisms, especially utilizing backprojections, are involved in building neurons with tuning, which is ethologically helpful, such as neurons which are useful in discriminating between faces, and other neurons which are useful in categorizing facial expressions (Rolls, 1987b, 1989a,b). This combination of multiple layers of competitive neuronal nets with some guidance provided by backprojections provides, we believe, one way to ensure that only useful filters are built, in order to circumvent the hazard which otherwise arises because of a combinatorial explosion (Rolls, 1989a,b).

SUMMARY

The ways in which information about faces is represented and stored in the temporal lobe visual areas of primates as shown by recordings from single neurons have been considered.

1. Neurons with responses which occur primarily to faces are found particularly in certain areas of the temporal lobe cortex, particularly in an area in the upper bank of the anterior part of the superior temporal sulcus (area TPO), and in areas TEa and TEm in the anterior part of the cortex forming the ventral lip of the superior temporal sulcus.
2. These neurons in many cases respond differently to the faces of different individuals, so that information about face identity which would be useful in recognition is represented by some of the neurons. However, the face of each individual is coded by the pattern of firing across a subpopulation of neurons. That is, ensemble encoding rather than "grandmother cell" encoding is used.

3. It was argued that the type of tuning found is a delicate compromise between very fine tuning, which has the advantage of low interference in neuronal network operations but the disadvantage of losing the emergent properties of storage in neuronal networks, and broad tuning, which has the advantage of allowing the emergent properties of neuronal networks to be realized but of leading to interference between the different memories stored in an associative network. Neurons in these areas are seen as filters which, as an ensemble, give a unique representation of a particular stimulus in the environment.

4. There is evidence that the responses of some of these neurons are altered by experience so that new stimuli become incorporated in the network. The role of competition in the function of such networks was considered.

5. It was shown that the representation which is built in temporal cortical areas shows considerable size, contrast, spatial frequency and translation invariance. Thus the representation is in a form which is particularly useful for storage and as an output from the visual system.

6. It was also shown that one of the representations which is built is object based rather than viewer centered.

7. In addition to the population of neurons which code for face identity, there is a separate population which conveys information about facial expression.

REFERENCES

Andersen, P.O. (1987). Properties of hippocampal synapses of importance for integration and memory. In G.M. Edelman, W.E. Gall, & W.M. Cowan, (Eds.), *New insights into synaptic function* (pp. 403–29). New York: Neuroscience Research Foundation/Wiley.

Azzopardi, P., & Rolls, E.T. (1989). Translation invariance in the responses of neurons in the inferior temporal visual cortex of the macaque. *Society for Neuroscience Abstracts, 15,* 120.

Barlow, H.B. (1972). Single units and sensation: a neuron doctrine for perceptual psychology? *Perception, 1,* 371–394.

Baylis, G.C., Rolls, E.T., & Leonard, C.M. (1985). Selectivity between faces in the responses of a population of neurons in the cortex in the superior temporal sulcus of the monkey. *Brain Research, 342,* 91–102.

Baylis, G.C., Rolls, E.T., & Leonard, C.M. (1987). Functional subdivisions of temporal lobe neocortex. *Journal of Neuroscience, 7,* 330–342.

Bliss, T.V.P., & Lomo, T. (1973). Long-lasting potentiation of synaptic transmission in the dentate area of the anaesthetized rabbit following stimulation of the perforant path. *Journal of Physiology, 232,* 331–356.

Bruce, C., Desimone, R., & Gross, C.G. (1981). Visual properties of neurons in a polysensory area in superior temporal sulcus of the macaque. *Journal of Neurophysiology, 46,* 369–384.

Bruning, J.C., & Kintz, B.L. (1977). *Computational handbook of statistics* (2nd ed.). Glenview, IL: Scott, Foresman.

Campbell, F.W. (1983). Why do we measure contrast sensitivity? *Behavioural Brain Research, 10*, 87–97.

Cotman, C.W., Monaghan, D.T., & Ganong, A.H. (1988). Excitatory amino acid neurotransmission: NMDA receptors and Hebb-type synaptic plasticity. *Annual Review of Neuroscience, 11*, 61–80.

Cowey, A. (1979). Cortical maps and visual perception. *Quarterly Journal of Experimental Psychology, 31*, 1–17.

Damasio, A.R., Damasio, H., & Van Hoesen, G.W. (1982). Prosopagnosia: Anatomic basis and behavioral mechanisms. *Neurology, 32*, 331–341.

Desimone, R., & Gross, C.G. (1979). Visual areas in the temporal lobe of the macaque. *Brain Research, 178*, 363–380.

Desimone, R., Albright, T.D., Gross, C.G., & Bruce, C. (1984). Stimulus-selective properties of inferior temporal neurons in the macaque. *Journal of Neuroscience, 4*, 2051–2062.

Egan, J.P. (1975). *Signal detection theory and ROC analysis.* New York: Academic Press.

Green, D.M., & Swets, J.A. (1966). *Signal detection theory and psychophysics.* New York: Wiley.

Gross, C.G., & Mishkin, M. (1977). The neural basis of stimulus equivalence across retinal translation. In S. Harnad R.W. Doty, L. Goldstein, J. Jaynes, & G. Hrouthamer (Eds.), *Lateralization in the nervous system* (pp. 109–122). New York: Academic Press.

Gross, C.G., Desimone, R., Albright, T.D., & Schwartz, E.L. (1985). Inferior temporal cortex and pattern recognition. *Experimental Brain Research, Supplement, 11*, 179–201.

Grossberg, S. (1987). Competitive learning: From interactive activation to adaptive resonance. *Cognitive Science, 11*, 23–63.

Hasselmo, M.E., Rolls, E.T., & Baylis, G.C. (1986a). Object-centered encoding of faces by neurons in the cortex in the superior temporal sulcus of the monkey. *Society for Neuroscience Abstracts, 11*, 1369.

Hasselmo, M.E., Rolls, E.T., & Baylis, G.C. (1986b). Selectivity between facial expressions in the responses of a population of neurons in the superior temporal sulcus of the monkey. *Neuroscience Letters, S26*, S571.

Hasselmo, M.E., Rolls, E.T., & Baylis, G.C. (1989). The role of expression and identity in the face-selective responses of neurons in the temporal visual cortex of the monkey. *Behavioural Brain Research, 32*, 203–218.

Hasselmo, M.E., Rolls, E.T., Baylis, G.C., & Nalwa,V. (1989). Object-centered encoding by face-selective neurons in the cortex in the superior temporal sulcus of the monkey. *Experimental Brain Research, 75*, 417–429.

Jordan, M.I. (1986). An introduction to linear algebra in parallel distributed processing. In D.E. Rumelhart & J.L. McClelland (Eds.), *Parallel distributed Processing, Vol. 1. Foundations* (pp. 365–442).Cambridge, MA: MIT Press.

Kelso, S.R., Ganong, A.H., & Brown, T.H. (1986). Hebbian synapses in the hippocampus. *Proceedings of the National Academy of Science, 83*, 5326–5330.

Kohonen, T. (1988). *Self-organization and associative memory* (2nd ed.). New York: Springer-Verlag.

Leonard, C.M., Rolls, E.T., Wilson, F.A.W., & Baylis, G.C. (1985). Neurons in the amygdala of the monkey with responses selective for faces. *Behavioural Brain Research, 15*, 159–176.

Levy, W.B. (1985). Associative changes in the synapse: LTP in the hippocampus. In W.B. Levy, J.A. Anderson, & L. Lehmkuhle (Eds.), *Synaptic modification, neuron selectivity, and nervous system organization* (pp. 5–33). Hillsdale, NJ: Erlbaum.

Levy, W.B., & Desmond, N.L. (1985). The rules of elemental synaptic plasticity. In W.B. Levy, J.A. Anderson, & S. Lehmkuhle (Eds.), *Synaptic modification, neuron selectivity, and nervous system organization* (pp. 105–121).Hillsdale, NJ: Erlbaum.

Marr, D. (1970). A theory for cerebral cortex. *Proceedings of the Royal Society, B, 176*, 161–234.

Marr, D. (1971). Simple memory: a theory for archicortex. *Philosophical Transactions of the Royal Society, B, 262*, 23–81.

Maunsell, J.H.R., & Newsome, W.T. (1987). Visual processing in monkey extrastriate cortex. *Annual Review of Neuroscience, 10*, 363–401.

McNaughton, B.L. (1984). Activity dependent modulation of hippocampal synaptic efficacy: some implications for memory processes. In W. Seifert (Ed.), *Neurobiology of the hippocampus* (pp. 233–252). London: Academic Press.

Perrett, D.I., Mistlin, A.J., & Chitty, A.J. (1987). Visual neurons responsive to faces. *Trends in Neurosciences, 10*, 358–364.

Perrett, D.I., Rolls, E.T., & Caan, W. (1982). Visual neurons responsive to faces in the monkey temporal cortex. *Experimental Brain Research, 47*, 329–342.

Perrett, D.I., Smith, P.A.J., Mistlin, A.J., Chitty, A.J., Head, A.S., Potter, D.D., Broennimann, R., Milner, A.D., & Jeeves, M.A. (1985a). Visual analysis of body movements by neurons in the temporal cortex of the macaque monkey: A preliminary report. *Behavioural Brain Research, 16*, 153–170.

Perrett, D.I., Smith, P.A.J., Potter, D.D., Mistlin, A.J., Head, A.S., Milner, D., & Jeeves, M.A. (1985b). Visual cells in temporal cortex sensitive to face view and gaze direction. *Proceedings of the Royal Society, 223B*, 293–317.

Pinker, S. (1984) Visual cognition: an introduction. *Cognition, 18*, 1–63.

Rolls, E.T. (1981a). Processing beyond the inferior temporal visual cortex related to feeding, learning, and striatal function. In Y. Kasuki, R. Norgren, & M. Sato (Eds.), *Brain mechanisms of sensation* (pp. 241–269). New York: Wiley.

Rolls, E.T. (1981b). Responses of amygdaloid neurons in the primate. In Y.Ben-Ari (Ed.), *The amygdaloid complex* (pp. 383–393). Amsterdam, Netherlands: Elsevier.

Rolls, E.T. (1984). Neurons in the cortex of the temporal lobe and in the amygdala of the monkey with responses selective for faces. *Human Neurobiology, 3*, 209–222.

Rolls, E.T. (1985). Connections, functions and dysfunctions of limbic structures, the prefrontal cortex, and hypothalamus. In M. Swash & C. Kennard (Eds.), *The scientific basis of clinical neurology* (pp. 201–213). London: Churchill Livingstone.

Rolls, E.T. (1986a). A theory of emotion, and its application to understanding the neural basis of emotion. In Y. Oomura (Ed.), *Emotions. neural and chemical control* (pp. 325–344). Tokyo: Japan Scientific Societies Press, and Basel: Karger.

Rolls, E.T. (1986b). Neural systems involved in emotion in primates. In R. Plutchik & H. Kellerman (Eds.), *Emotion: Theory, research, and experience. Vol. 3: Biological Foundations of Emotion*, New York: Academic Press.

Rolls, E.T. (1986c). Neuronal activity related to the control of feeding. In R.C. Ritter, S.

Ritter, & C.D. Barnes (Eds.), *Feeding behavior: Neural and humoral controls* (pp. 163–190). New York: Academic.

Rolls, E.T. (1987a). Information representation, processing and storage in the brain: analysis at the single neuron level. In J.-P. Changeux & M. Konishi (Eds.), *The neural and molecular bases of learning* (pp. 503–540). Chichester: Wiley.

Rolls, E.T. (1987b). A neurophysiological systems approach to neuroethology. In D.M. Guthrie (Ed.), *Aims and methods in neuroethology* (pp. 231–259). Manchester, England: Manchester University Press.

Rolls, E.T. (1989a). Functions of neuronal networks in the hippocampus and neocortex in memory. In J.H. Byrne & W.O. Berry (Eds.), *Neural models of plasticity: Experimental and theoretical approaches* (pp. 240–265). San Diego, CA: Academic Press.

Rolls, E.T. (1989b). The representation and storage of information in neuronal networks in the primate cerebral cortex and hippocampus. In R. Durbin, C. Miall, & G. Mitchison (Eds.), *The computing neuron* (pp. 125–129). Wokingham, England: Addison-Wesley.

Rolls, E.T. (1989c). Functions of the primate hippocampus in spatial processing and memory. In D.S. Olton & R.P. Kesner (Eds.), *Neurobiology of comparative cognition* (pp. 339–362), Hillsdale, NJ: Erlbaum.

Rolls, E.T. (1989d). Information processing in the taste system of primates. *Journal of Experimental Biology, 65,* 141–164.

Rolls, E.T. (1989e). Functions of neuronal networks in the hippocampus and cerebral cortex in memory. In R.M.J. Cotterill (Ed.), *Models of brain function.* Cambridge: Cambridge University Press.

Rolls, E.T. (1990). A theory of emotion, and its application to understanding the neural basis of emotion. *Cognition and Emotion, 4,* 161–190.

Rolls, E.T., & Baylis, G.C. (1986). Size and contrast have only small effects on the responses to faces of neurons in the cortex of the superior temporal sulcus of the monkey. *Experimental Brain Research, 65,* 38–48.

Rolls, E.T., Baylis, G.C., & Hasselmo, M.E. (1987). The responses of neurons in the cortex in the superior temporal sulcus of the monkey to band-pass spatial frequency filtered faces. *Vision Research, 27,* 311–326.

Rolls, E.T., Baylis, G.C., Hasselmo, M.E., & Nalwa, V. (1989). The effect of learning on the face-selective responses of neurons in the cortex in the superior temporal sulcus of the monkey. *Experimental Brain Research, 76,* 153–164.

Rolls, E.T., Baylis, G.C., & Leonard, C.M. (1985). Role of low and high spatial frequencies in the face-selective responses of neurons in the cortex in the superior temporal sulcus. *Vision Research, 25,* 1021–1035.

Rolls, E.T., Thorpe, S.J., Boytim, M., Szabo, I., & Perrett, D.I. (1984). Responses of striatal neurons in the behaving monkey. 3. Effects of iontophoretically applied dopamine on normal responsiveness. *Neuroscience, 12,* 1201–1212.

Rolls, E.T. & Williams, G.V. (1987). Sensory and movement-related neuronal activity in different regions of the primate striatum. In J.S. Schneider & T.I. Lidsky (Eds.), *Basal ganglia and behavior: Sensory aspects and motor functioning.* Bern: Hans Huber.

Rolls, E.T., & Treves, A. (1990). The relative advantages of sparse versus distributed encoding for associative neuronal networks in the brain. *Network,* in press.

Rumelhart, D.E., & McClelland, J.L. (1986). *Parallel distributed processing.* Cambridge, MA: MIT Press.

Rumelhart, D.E., & Zipser, D. (1985). Feature discovery by competitive learning. *Cognitive Science, 9,* 75–112.

Seltzer, B., & Pandya, D.N. (1978). Afferent cortical connections and architectonics of the superior temporal sulcus and surrounding cortex in the rhesus monkey. *Brain Research, 149,* 1–24.

Smith, D.V., & Travers, J.B. (1979). A metric for the breadth of tuning of gustatory neurons. *Chemical Senses, 4,* 215–229.

Wigstrom, H., Gustaffson, B., Huang, Y.-Y., & Abraham, W.C. (1986). Hippocampal long-term potentiation is induced by pairing single afferent volleys with intracellularly injected depolarizing currents. *Acta Physiologica Scandinavia, 126, 317–319.*

4

Towards a Quantitative Understanding of Facial Caricatures*

P.J. Benson
D.I. Perrett
Department of Psychology
　　University of St. Andrews
　　St. Andrews, Fife

D.N. Davis
Department of Medical Biophysics
　　University of Manchester
　　Manchester, England

INTRODUCTION

Facial caricaturing provides a vehicle for the expression of public and intimate sentiment. The caricaturist's results are often amusing and are an important part of the repertoire of entertainment, sometimes a component of satirical political comment. The illustrator's ability to make such a statement about a person's character by selectively incorporating inaccuracies into a portrait through distortion and exclusion of features is an ingenious and intriguing one (Figure 4.1). Whether the caricature is presented as a 2-D sketch or as a 3-D model, such as Central Television's notorious "Spitting Image" puppets, there are qualities in

---

* This work forms part of a collaborative investigation into face recognition funded by an ESRC Programme Award XC15250001 to Vicki Bruce (University of Nottingham), XC15250002 to Ian Craw (University of Aberdeen), XC15250003 to Hadyn Ellis (University of Wales College of Cardiff), XC15250004 to Andrew Ellis (University of York) and Andrew Young (University of Durham), and XC15250005 to David Perrett (University of St. Andrews). Phil Benson is funded by the ESRC St. Andrews award (XC15250005). David Perrett is a Royal Society University Research Fellow. Darryl Davis was funded by the MRC (G8427112N) at St. Andrews.

Our thanks go to Mr. Parsons for agreeing to be caricatured and permitting us to use the photographs in this chapter.

caricature which fuel our imaginations and may reveal much about the way in which we code faces in memory.

This chapter makes a survey of studies of the perception and recognition of line-drawn facial caricatures. We go on to focus on more recent attempts to produce caricatures without the skills of an artist through image processing techniques. The long-term aim of our work is to glean from the principles of caricature production and perception insights into how faces are normally coded in memory.

It has become a topic suitable for psychological investigation to try to determine how we are so easily able to identify faces from impoverished and distorted representations, even when the faces depicted are not entirely familiar. In attempting to answer how caricatures work, one can ask subsidiary questions:

- How is a caricature produced?
- What aspects of an image should be enhanced or omitted?
- How does the recognition of caricatures compare with the recognition of veridical images?

THE PARAMETERS OF CARICATURE

Perkins's (1975) attempt to define more clearly a *caricature*, highlights a number of aspects which are considered by the artist:

> a symbol referring to an individual and relative to a whole set of scales, norms, and populations is a caricature just when it is a caricature with respect to some of those scales, norms, and populations and accurate with respect to the others. (p. 7)

He suggests that *exaggeration* through *individuation* is of great importance; the subject's facial physiognomy is studied, and those features subjectively considered to be characteristic of that individual will more often than not be exaggerated.

Facial Features and Configurational Cues

In many cases it is possible that a single aspect of an image is sufficient to trigger recognition, for example, Nixon's nose (or hairline or jowls), but it is unlikely that, if the rest of the face was inappropriate, recognition from a single feature would be possible.

The appearance of individual facial features, and the spatial relationship between facial features, are both important dimensions in coding the identity of a face (see Haig, 1984, 1986; Rhodes, 1988); the caricature artist must incorporate both aspects to make a good rendition of a face. Other, less outstanding and memorable, attributes of the target face will be omitted or at least minimized. These omissions will be, by definition, components and a topographic relationship which the caricaturist considers ordinary or closest to a representation of an average face. So the facial features of the subject which the cartoonist thinks make the face distinguishable from other people will be best represented in the product. It is also possible to introduce mannerisms and personality traits into the caricature by manipulating the head posture, for example, Thatcher looking down her nose to symbolize aloofness (see Figure 4.1).

Paraphernalia

Sometimes we find environmental ingredients (people, buildings) and more often personal accoutrements (pipe, hat, spectacles) incorporated into the caricature image, much in the same way that personal image-makers mould a public appearance for a politician, for example, Harold Wilson's pipe, designed to contain extravagant hand-waving and to add a touch of homeliness to his personality during his election campaign prior to becoming Prime Minister.

Figure 4.1. Handdrawn caricature of Rt. Hon. Margaret Thatcher M.P.

Degree of Exaggeration

Having decided *what* to caricature, the degree of exaggeration must be decided. This will vary with the artist's epigrammatical style, bias, and the political and social climate (Perkins, 1975; Goldman & Hagen, 1978).

Goldman and Hagen's (1978) study of caricatures produced by 17 artists of Richard Nixon during the period 1972–1973 showed a very high concordance across artists and time as to *what* was caricatured. The degree of exaggeration relative to Nixon's actual face (assessed from measurements of five photographs) was, however, variable across both parameters, due to artistic style and bias; the degree of feature exaggeration varied across artists from 12 to 86 percent (mean = 56%). Physiognomic changes were present in the real face during the years of measurement, but these were less dramatic than the increases in feature distortion in the caricatures. In the evolution of the exaggerated style we are perhaps witnessing reactions to the onset of Watergate (and negative political climate) and/or the attempt by artists to suggest changes in Nixon's manner, that is, looking older, more tense, and strained because of the crisis.

Symbolic Reference

Famous faces and those currently in the public eye and consequently exposed to frequent portraiture may often require less artistic effort in depiction than does the caricaturing of less well-known characters (see Gombrich, 1969). In the latter, greater attention to detail would be required to help the reader/viewer, because the conventions of representation for that individual have not yet become agreed. That is, we may come to recognize the form of a Margaret Thatcher cartoon itself rather than perceiving it as a depiction of the real person. This may be more a matter of recognizing the symbolic reference of the drawing rather than recognizing the drawing for what it is. Indeed it is often remarked that people can recognize Spitting Image models as person X but cannot recognize original pictures or film of X.

To summarize, a facial caricature can be defined as a selective exaggeration of specific features which the artist—or convention—deems to be characteristic. These can be facial features, posture, expression, or associated paraphernalia.

HOW DO CARICATURES SUCCEED?

When we learn the appearance of a face, we are more likely to code the unusual or striking features. It will be the strength and accuracy of this encoding which enables recall of this particular face (Valentine & Bruce, 1986a,b; the *distinctiveness hypothesis*, Rhodes, Brennan, & Carey, 1987). Therefore, faces with highly distinctive features are recognized more efficiently than faces that lack distinctive features, which consequently appear fairly average in all dimensions (Valentine & Bruce, 1986a,b). Gibson (1971), Hagen (1974), and Hagen and Perkins (1983) have considered that the caricature sketch may contain an essential minimum of information for building an adequate face topology.

Precise facial detail is not necessary even for important features, and unimportant features may be completely ignored, so recognition may be initiated by partial information. Because there are so few key features to code, recognition will be inhibited by a caricature displaying misdirected exaggeration of an essential feature (using Perkins's 1975 terminology, a *contraindication* of these constituents). The information in a caricature may constitute, not simply descriptive fidelity, since the omitted information is redundant, but perhaps a *superfidelity*, since the remaining information is exaggerated. Caricatures could be considered in an ethological sense as supernormal stimuli, being more potent than the stimuli they represent (Hinde, 1982). For example, a gull will retrieve an artificial egg larger than its own (a supernormal stimulus) in preference to an egg of normal size.

Information Redundancy in Face Images

The classic edge-detection mechanism of Marr and Hildreth (1980) fails to "find" the perceptually important edges of a face (Pearson & Robinson, 1985; Pearson, Hanna, & Martinez, 1990). In contrast, the latter authors made use of an understanding of facial surfaces to design a local operator, a luminance *valley detector*, which, when convolved with a facial image, produces an effective line drawing representation. More importantly, with the valley detector it is possible to selectively mark edges that viewers perceive as important. That is, a face processed by the valley detector looks very much like that which is produced by a caricature artist without the exaggeration.

With further processing it has been possible for Pearson's group to produce highly realistic real-time monochrome images of actors moving, demonstrating electronically that very little information is required to define an image of a face which can be reliably perceived.

Caricature and Veridical Representations

While precise details may not be necessary for recognition, it is the case that photographs produce better recognition than veridical line-drawings. Davies, Ellis, and Shepherd (1978) demonstrated that photographs were more accurately recognized (89.7%) than were detailed line-drawings (46.8%) or outlines (23.5%). Perkins (1975), however, suggested that even a line-tracing of a photographed face does not prove as recognizable as a line-drawing caricature. The order of goodness of representation would thus appear to be photographs followed by caricatures followed by line drawings (detailed or simple), though the claim that caricatures are better representations than line drawings is contentious (see discussion of the caricature advantage below).

Psychological experimentation has predominantly used handdrawn line caricatures for stimuli. The efficacy of caricature representations has been assessed with ratings of likeness of the representation, familiarity and name–face matching. In some of these experiments two different stimuli media have been used (Hagen & Perkins, 1983; Tversky & Baratz, 1985; see also Bruce, 1983, 1988). That is, photographs have been compared with line-drawing caricatures. In these experiments the caricatures were not found to produce better recognition than veridical representations. Such a comparison is perhaps unlikely to demonstrate an advantage for the caricature given the findings of Davies et al. (1978) that the face photographs are recognized so much better than line drawings. Even though much of the photographic detail may be redundant, the fact that real photographs are better recognized implies that photographs still contain more nonredundant information than line drawings. It is almost certain that well-known faces will have already been witnessed in real life, photographs, film, or television (Nelson, Metzler, & Reed, 1974; Chance, Goldstein, & McBride, 1975). We can therefore assume that subjects will have learnt many details of familiar faces present only in high-resolution images.

Ryan and Schwartz (1956) and Hochberg (1978) report recognition responses to various representations of the same objects (including switches, hands, and valves) where cartoon caricatures were recognized more quickly than photographs or shaded line drawings, which in turn were better than line drawings (see Biederman, 1985, 1987, for a review). In the case of hands, photographs were best recognized, with shaded drawings second and cartoons third. Thus, when photographs are pitted against the drawings or line-drawn caricatures, the latter do not show any consistent advantage in supporting recognition.

GENERATING CARICATURES AUTOMATICALLY

The brief review above reveals that a general principle of facial caricaturing is to define features of the face that typify an individual, and then to exaggerate these. One can derive an algorithm for achieving this automatically:

> Compare the feature dimensions with the average for many faces, and, where differences are found, exaggerate these.

In the implementation of such an algorithm, it is not necessary to identify which particular features might be the most characteristic of a particular face, since, by exaggerating all feature differences, those of particular importance for recognition will automatically be included. It could be argued that, in caricatures produced by artists, it is also important to ignore nonsalient features. This too can be incorporated into the algorithm, by relating the degree of exaggeration to the degree a feature dimension departs from the normal. By definition, non-salient features should be close to normal.

Brennan (1985) followed this approach and developed a computer program for automatically producing line-drawing caricature portraits of frontal faces. Digitized images of faces had features manually highlighted by logging 169 feature points which allowed the face to be represented by 37 lines, for example, hair, eyes, nose, mouth, ears. By systematic comparison of these x-y coordinates of each point on a target face with those of a norm or average, all differences could be exaggerated to produce a caricature of the target face. For example a nose which is 20 units longer than the associated norm will be increased by 5 units in a caricature which represents a 25 percent exaggeration from the original face. Similarly, *anticaricatures* could be derived by decreasing the difference between the face and the norm; a −25 percent anticaricature can be produced by depicting the nose as 15 units in length.

A 50 percent caricature is therefore defined as a picture in which the distances of all the feature points from the corresponding positions in the normal face have been increased by half as much again. Similarly, a 100 percent feature caricature will be twice as far away from the norm.

The "Average" Face

For an automatically produced caricature to be effective, there must be a good notion of an average or **base** face against which the target face is to be contrasted. Normally the artist would compare a face with an average of comparable age, sex and race. However, if an attempt is being made to deride a politician, then he or she may be exaggerated towards a young child, or even an animal (e.g., donkey, parrot). Other occasions may suggest comparisons across sex or race. A person's occupation may even suggest how to caricature his or her face, since the relation between face and occupation is not entirely arbitrary; for example, politicians tend to look old and have well-groomed short hair, pop stars tend to look young and have long hair, and so on. (See Cross, Cross, & Daly, 1971; Dion, Breschied, & Walster, 1972; Klatzky, Martin, & Kane, 1982a,b; and Klatzky & Forrest, 1984, for a discussion.)

It is clear that different caricatures should be generated using different base faces since variations in age, and differences in sex and race, for this starting point could have wide-ranging effects. Chance et al. (1975) and Shepherd (1981) demonstrated that people recognize faces from their own race better than faces of others. Cross-racial caricaturing (where a face is contrasted with the norm of some other race) in psychological investigation could be considered undesirable unless such specific effects are being examined.

Rhodes et al. (1987) used a selection of averaged faces on which to base their caricatures; four average faces (norms), two of each sex, were used. One of each sex was derived by averaging three faces, male or female, previously rated as typical (av3); the remaining two were generated using either 10 female or 8 male faces (avpsych).

Dewdney (1986) produced a very naïve model of an average *androgynous* face by selecting some features from Rhodes et al.'s (1987) average male face and some from the average female face. Dewdney's 'face' does not represent a true average model. For this one would need to derive the mean of the positions of each feature from both male and female averages.

In these latter two cases, x-y coordinates have been used to represent the average face data, which facilitates simple comparison of feature locations in "face-space."

A Caricature Advantage with Computer Caricatures?

Rhodes et al. (1987) used the computational procedures of Brennan (1985) to help perform a psychological investigation of recognition with quantitatively specified caricatures. Rhodes et al. compared judgments of familiar faces depicted as veridical line drawings or as caricatures, overcoming the problem of cross-media comparisons.

Student subjects recognized caricatures (+50% exaggeration) of departmental staff and students more quickly than the veridical representations (0% distortion), which were in turn recognized more quickly than the anticaricatures (−50%). Pooled mean ± S.E. reaction times (in seconds) for recognition of the target faces depicted were 12.3 ± 2.3 (−50%), 6.4 ± 1.3 (0%) and 3.2 ± 0.4 (+50%). Although the caricatures were recognized more quickly than the veridicals, they were identified less accurately, though the difference in accuracy did not reach significance in this experiment (mean proportion of correct identifications were 33 percent (+50%), 38 percent (0%) and 27 percent (−50%). Further, veridical representations were recognized more quickly than anticaricatures but not more accurately. Thus, from reaction-time data (but not from accuracy data) there appears to be a caricature advantage. Indeed a speed/accuracy tradeoff (faster recognition but lower accuracy) has also been found in a more recent comparison between caricatures of veridical line-drawings of famous faces (S. Carey, personal communication 1989). This speed/accuracy trade off complicates any interpretation of a supposed caricature advantage.

While the above focused on subjects' ability to recognize faces, Rhodes et al. (1987) also investigated how well line drawings and caricatures were judged to represent a target face. A likeness test was administered using a set of 7 stimuli (−75%, −50%, −25%, 0% (veridical), +25%, +50% and +75% caricatures), randomly ordered. Subjects were asked to rate the goodness of likeness of each stimuli on a scale from 1 = poor to 7 = great (mean rating = 5.1). Highest ratings were found for 0 percent and +25 percent (of approximately equal mean score).

By fitting a curve to the data, relating likeness to degree of caricature, Rhodes et al. found that an interpolated figure of approximately +16 percent caricature would produce the highest likeness ratings. The data show that the likeness value of such an image is unlikely to be significantly different from the mean rating for 0 percent and +25 percent caricatures.

The results from the investigations with computer-generated line-drawing caricatures are important because they indicate that a *systematic* distortion of an image can lead to better recognition and perhaps a better representation than a veridical image. The result has implications for the nature of representations stored in memory.

Two explanations have been advanced to account for the caricature advantage concerning mental representation of faces:

1. That we store a caricature representation rather than a veridical representation of a face; or
2. That we store a veridical representation, but that the caricature somehow constrains the search space in our attempts to match the input image to the stored descriptions or templates of faces.

The second alternative might occur if faces are stored as *distances* in multidimensional space at the center of which is the norm for a particular face type (McClelland & Rumelhart, 1985; Valentine & Bruce, 1986b; Rhodes, 1988; Valentine, 1988). Thus, nose length might be one dimension, interocular separation another.

When a caricature is perceived, the exaggeration of particular features will increase the distance of the caricatured face from representations of other faces. Hence, one can speculate that an advantage for the caricature should result from an increased ease of matching the caricature to the veridical representation of the target face; that is, caricatures optimize the retrieval process rather than mimic the storage process. The only problem with this interpretation is that a caricature image will also be further away from the template of the target face (compared to the veridical image). The caricature advantage must come from the fact that, in multidimensional space, the small increase in distance from the target face is more than offset by the large increase in distance from nontarget faces.

Significantly, Rhodes et al. report that there was no evidence to suggest that the type of norm, *av3* or *avpsych*, affected the identifiability of the caricatures; the "best" drawing of the target faces based on these averages was calculated to be +16 percent, and +15 percent, respectively. This could be interpreted as suggesting that the average against which a face is compared for caricaturing is unimportant (in contrast to the "Average Face" section), but we note that the norms with which comparisons were made are very similar.

The distinctiveness hypothesis ("Generating Caricatures Automatically") was supported by the fact that caricatures of every degree (+25%, +50%, +75%) were judged better likenesses than the anticaricatures of the same degree. Therefore, exaggeration of the distinctive aspects of a face (features most unlike those in a norm) synthesizes a better likeness than a decrease by the same amount. A caricature advantage over veridical representation in speed of recognition is clear from the work of Rhodes et al., although previous authors had not been able to show this (Hagen & Perkins, 1983; Tversky & Baratz, 1985), perhaps because a scientific model for generating caricatures was not available.

Rhodes et al. found no caricature advantage for the recognition of unfamiliar faces, suggesting that the first model in the representation hypothesis (where we would store an exaggerated representation of a face in long-term memory) was more likely to be correct. The second model (where a caricature gives faster access to a veridical coding) is perhaps favored from the likeness ratings for familiar faces, because +50 percent caricatures were recognized faster but were found to be no better likenesses than the veridical stimuli.

EXTENDING COMPUTER CARICATURES TO PHOTOGRAPHIC IMAGES

We have extended Brennan's caricature process to allow increased flexibility in the feature logging and caricature generation phases. Additionally we can render the resulting caricatures as full-color photographic images (Benson & Perrett, 1991a).

Studies so far have only been performed with line drawings. Thus the caricature effect may only tell us about the perception of cartoon conventions and little about the recognition of real images. Furthermore, the advantage of caricature to date, as detected by Rhodes et al. (1987), is only small, particularly for judgments on the likeness of a representation to the original person.

We can now use these tools to assess:

1. Whether the advantage of caricatures is restricted to line-drawings, and
2. Whether it is possible to create "superfidelity" images of people which look more like them than natural images do; since photographs are better representations than line-drawn caricatures, it is not possible to create superfidelity with line drawings.

Procedures for Interactive Synthesis

Twenty-four-bit color images (or 8-bit grey scale) are frame-grabbed on a Pluto 24i[1] system and transferred to an IRIS[2] graphics processor. One hundred eighty-six points are selected on the source image interactively and selected points

---

[1] Pluto is a registered trademark of Io Research Ltd.
[2] IRIS is a registered trademark of Silicon Graphics, Inc.

joined to form 37 lines delineating the features of the face (Figure 4.2). The feature coordinates are added to a database of existing faces which can be selected for manipulation. Distortion of the face image (Figure 4.3) is interactively guided in real-time by accentuating differences from an average face data set. Exaggerations are performed using vector deformation by manipulating all metric distances between the points of the average face and those of the subject face.

The resulting feature coordinates are used to map the original image onto its caricature counterpart. This is achieved by triangulating the original and caricature faces (Figure 4.4); the vertices of each tessellation lie at the feature points specified at the previous stage. The grey scale or color contents of each triangle can then be remapped pixel by pixel from the original image to the caricature image, according to the geometric distortion between source and destination triangles. The resulting image appears as a smoothly transformed face without any fractures along the edges of the tessellations. Figures 4.5, 4.6, and 4.7 show such caricatures.

It is not feasible to rely on an automatic triangulation function to generate the necessary tessellations of the face, as this may cause inappropriate overlapping of polygons (overlapping usually occurs at > +50%). Instead, a static array of 328 triangular tessellations are specified over the image. The array is interactively accessible to permit overlaying of specific portions of the image, for

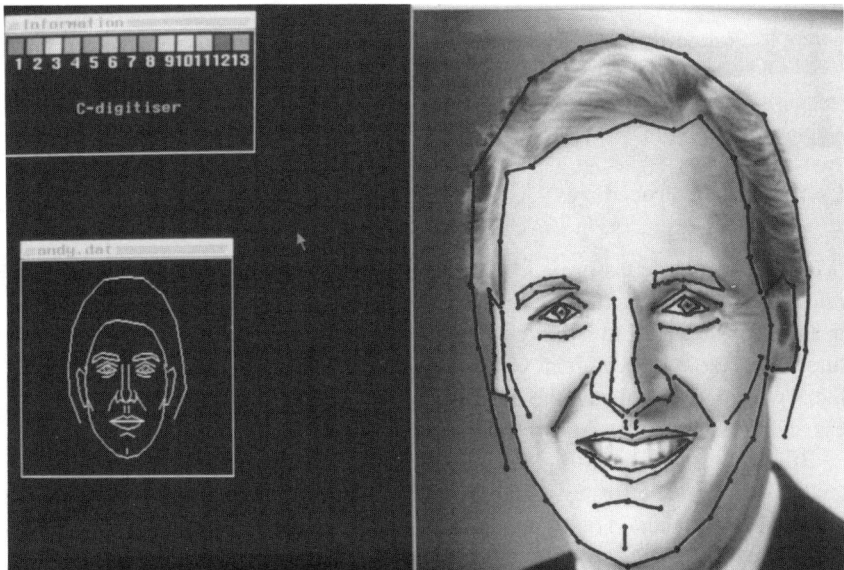

Figure 4.2. A digitized image of Mr. Nicholas Parsons overlaid with 186 feature points. The smaller face panel shows an average key whose features are color-coded in sequence indicating to the user the approximate locations of each of the points.

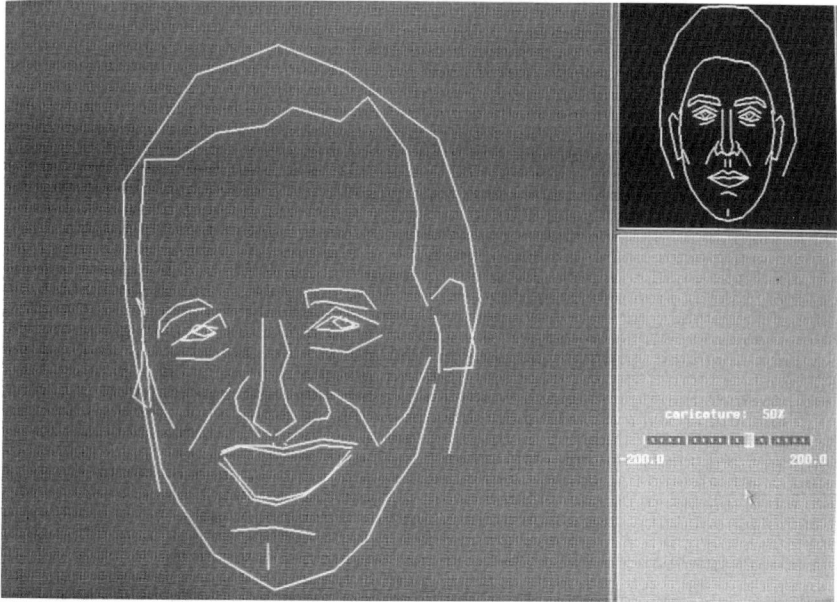

Figure 4.3. A +50% exaggeration of the veridical data for the digitized image of Nicholas Parsons. Lines link adjacent points around facial features in a manner similar to those of Brennan (1985).

example, where the eyes and eyebrows are made to remain visible when they would be obscured by overlapping forehead or hair tessellations in the caricature image by rendering the eye tessellations last of all.

Comments on Procedures

Scale and *rotate* transformations can be applied to each face to ensure that interpupilary distances are constant. People whose posture naturally dictates a head tilt or bow will thus be caricatured differently. To produce the best caricature of a person, the veridical Cartesian data should capture the verisimilitude of their physiognomy using the limited set of feature points available. This is especially true for well-known public faces.

Bruce, Valentine, and Baddeley (1987) note a coding preference for a three-quarter view profile for heads/faces. Most handdrawn caricatures are portrayed adopting this stance. This is modestly reflected in the generation phase of our caricatures, as a slight rotation of the head around the y-axis will cause greater distortions of such key features as the nose and chin than would be perceived if the face looks straight at the camera. Rotation of the head in the image must not be so great as to prevent digitization of hidden features, for example, ear, eye,

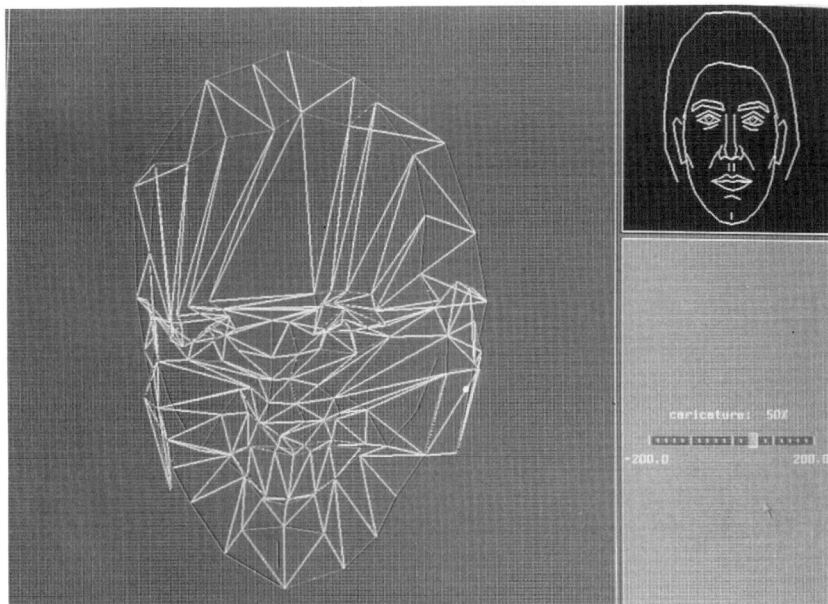

Figure 4.4. Feature points data from +50% caricature (Figure 4.3) are remapped into 328 triangles. For clarity, tessellations between the face image and the picture border are not shown. The tessellated image may also be interactively distorted to facilitate checking for overlapping triangulations which can cause anomalies in the subsequent production of a photographic caricature.

hairlines. Only by using run-time generation or modification of the tessellation map could features not visible be omitted. Features such as ears, which are often occluded by hair, must be subjectively highlighted at the delineation stage. Furthermore, face attitude and facial expression captured at some point in the process will produce the greatest exaggerations in caricature according to an *upright* norm displaying a neutral expression.

Metric distortions will not synthesize caricatures in the same way as an artist's sketch. Firstly, hand-drawings are free to incorporate minor (or major) deviances from such a metrical theory, perhaps to capture a particular mood or physical nuance. The computer model always attempts to make exaggerations. The second reason lies with the choice of norm. Different caricatures of the same person will be generated by a comparison with averages of differing age, sex and race; this has greatest implications for the anticaricatures. Efficacious caricaturing therefore relies on a careful selection of the model's parameters. Additionally, one can only produce caricatures of the face captured in a particular photograph, which may not necessarily be the best likeness of that person.

The possibility remains to use a caricature artist's skills when the distortions are being made, so adjusting the degree to which particular features are exagger-

Figure 4.5. A photographic caricature (rated at +50%) of Nicholas Parsons, produced by mapping the pixel values within each of the triangles of the original tessellation map (0% exaggeration) onto the equivalent, but distorted, triangles in the +50% map.

ated. Although this is more in line with traditional cartooning, it does not in itself help us understanding caricatures.

Judgments of Likeness with Caricature Photographs

We have investigated 30 subjects' estimate of "best likeness" for normal, +ve caricature, and −ve caricature grey-scale images of seven faces (Benson &

Figure 4.6. The original (0% exaggeration) image of Nicholas Parsons.

Perrett, 1991b). The faces were chosen from well-known TV personalities, six of whom are not known to have been subjected to caricaturing in the media. Figure 4.8 gives a summary of the results in the study. We found that subjects consistently rate +ve caricatures and normal images as more like the person they represent than −ve caricatures. There was, however, no significant difference between +ve caricatures and normal faces. The mean of the distribution as a whole shows a small but significant ($t = 3.6$, df $= 5$, $p < 0.05$) shift to the +ve caricaturing (this was true for six of the seven faces examined individually).

Figure 4.7. An anticaricature (rated at −50%) of Nicholas Parsons, produced as Figure 4.5 but with a tessellation map distorted 50% towards the androgynous face.

Interpolation (using a polynomial curve) would suggest that a +9 percent caricature would produce the best likeness.

The study does not lend strong support for a caricature advantage, since the +ve caricatures (+16%) were not found to be better likenesses than veridical (0%) faces. The results are in fact very similar to those of Rhodes et al. (1987) for the comparable perceptual task of rating "best likeness." Their study showed very little difference between 0% (veridical) and +25 percent caricature.

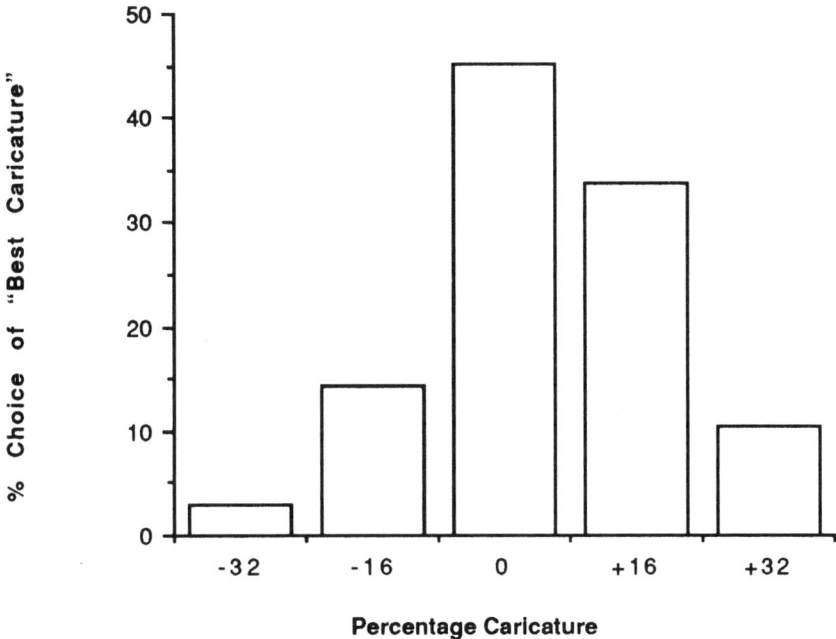

Percentage Caricature

Figure 4.8. Percentage caricature plotted against percentage sample choice of "best likeness." Subjects were shown −32%, −16%, 0%, +16%, and +32% caricatures. Although there is no definite caricature advantage, the data do show a preference for caricatures over anticaricatures.

We are at present involved in a study assessing efficiency of recognition of normal and +ve caricatures. It is still possible that, like the Rhodes et al. study, this measure would give us a caricature advantage, particularly with high-exposure faces.

In conclusion, the procedures for creating caricatures from photographic images have thus been established to a large extent. The technique not only has an intrinsic amusement value, but also provides a tool for psychology which promises to reveal more about the nature of memory representations for faces.

REFERENCES

Benson, P.J., & Perrett, D.I. (1991a). Synthesising continuous-tone caricatures. *Image & Vision Computing*.

Benson, P.J., & Perrett, D.I. (1991b). Perception & recognition of photographic quality facial caricatures: Implications for the recognition of natural images. *European Journal of Cognitive Psychology*.

Biederman, I. (1985). Human image understanding: Recent research and a theory. *Computer Vision, Graphics & Image Processing, 32*, 29–73.

Biederman, I. (1987). Recognition-by-components: A theory of human image understanding. *Psychological Review*, *94*(2), 115–147.

Brennan, S.E. (1985). Caricature generator: Dynamic exaggeration of faces by computer. *Leonardo*, *18*(3), 170–178.

Bruce, V. (1983). Recognising faces. *Philosophical Transactions of The Royal Society of London*, *B302*, 423–436.

Bruce, V. (1988). *Recognising faces*. London: Erlbaum.

Bruce, V., Valentine, T., & Baddeley, A. (1987). The basis of the 3/4 view advantage in face recognition. *Applied Cognitive Psychology*, *1*, 109–120.

Chance, J., Goldstein, A.G., & McBride, L. (1975). Differential experience and recognition memory for faces. *Journal of Social Psychology*, *97*, 243–253.

Cross, J.F., Cross, J., & Daly, J. (1971). Sex, race, age & beauty as factors in recognition of faces. *Perception & Psychophysics*, *10*, 393–396.

Davies, G.M., Ellis, H.D., & Shepherd, J.W. (1978). Face recognition accuracy as a function of mode of representation. *Journal of Applied Psychology*, *63*, 180–187.

Dewdney, K.A. (1986). Computer recreations: The compleat computer caricaturist and a whimsical tour of face space. *Scientific American*, *255*, 20–28.

Dion, K.K., Brescheid, E., & Walster, E. (1972). What is beautiful is good. *Journal of Personality & Social Psychology*, *24*(3), 285–290.

Gibson, J.J. (1971). The information available in pictures. *Leonardo*, *4*, 27–35.

Goldman, M., & Hagen, M.A. (1978). The forms of caricature: Physiognomy and political bias. *Studies in the Anthropology of Visual Communication*, *5*, 30–36.

Gombrich, E.H. (1969). *Art and illusion: A study in the psychology of pictorial representation*. Princeton, NJ: Princeton University Press.

Hagen, M.A. (1974). Picture perception: Toward a theoretical model. *Psychological Bulletin*, *81*(8), 471–497.

Hagen, M.A., & Perkins, D. (1983). A refutation of the hypothesis of the superfidelity of caricatures relative to photographs. *Perception*, *12*, 55–61.

Haig, N.D. (1984). The effect of feature displacement on face recognition. *Perception*, *13*, 505–512.

Haig, N.D. (1986). High-resolution facial feature saliency mapping. *Perception*, *15*, 373–386.

Hinde, R.A. (1982). *Ethology: Its nature and relations with other sciences*. Glasgow, Scotland: Collins.

Hochberg, J. (1978). *Perception* (2nd ed.). Englewood Cliffs, NJ: Prentice-Hall.

Klatzky, R.L., Martin, G.L., & Kane, R.A. (1982a). Influence of social-category activation on processing of visual information. *Social Cognition*, *1*, 95–109.

Klatzky, R.L., Martin, G.L., & Kane, R.A. (1982b). Semantic interpretation effects on memory for faces. *Memory & Cognition*, *10*(3), 195–206.

Klatzky, R.L., & Forrest, F.H. (1984). Recognizing familiar and unfamiliar faces. *Memory & Cognition*, *12*(1), 60–70.

McClelland, J.L., & Rumelhart, D.E. (1985). Distributed memory and the representation of general and specific information. *Journal of Experimental Psychology: General*, *114*(2), 159–188.

Marr, D., & Hildreth, E. (1980). Theory of edge detection. *Proceedings of the Royal Society of London*, *B207*, 187–217.

Nelson, T.O., Metzler, J., & Reed, D.A. (1974). Role of details in the long-term recognition of pictures and verbal descriptions. *Journal of Experimental Psychology*, *102*(1), 184–186.

Pearson, D.E., Hanna, E., & Martinez, K. (1990). Computer-generated cartoons. In H. Barlow, C. Blakemore, & M. Weston-Smith (Eds.), *Images and understanding*. Cambridge, England: Cambridge University Press.

Pearson, D.E., & Robinson, J.A. (1985). Visual communication at very low data rates. *Proceedings of the IEEE*, *73*(4), 795–812.

Perkins, D. (1975). A definition of caricature and recognition. *Studies in the Anthropology of Visual Communication*, *2*(1), 1–24.

Rhodes, G., Brennan, S., & Carey, S. (1987). Identification and ratings of caricatures: Implications for mental representations of faces. *Cognitive Psychology*, *19*, 473–497.

Rhodes, G. (1988). Looking at faces: First-order and second-order features as determinants of facial appearance. *Perception*, *17*, 43–63.

Ryan, T.A., & Schwartz, C.B. (1956). Speed of perception as a function of mode of representation. *American Journal of Psychology*, *69*, 60–69.

Shepherd, J.W. (1981). Social factors in face recognition. In G.M. Davies, H.D. Ellis, & J.W. Shepherd (Eds.), *Perceiving and remembering faces*. New York: Academic Press.

Tversky, B., & Baratz, D. (1985). Memory for faces: Are caricatures better than photographs? *Memory & Cognition*, *13*(1), 45–49.

Valentine, T. (1988, February). Report on current research to the 5th Grange-over-Sands meeting on Face Perception.

Valentine, T., & Bruce, V. (1986a). Recognising familiar faces: The role of distinctiveness and familiarity. *Canadian Journal of Psychology*, *40*(3), 300–305.

Valentine, T., & Bruce, V. (1986b). The effects of distinctiveness in recognising and classifying faces. *Perception*, *15*, 525–535.

5
Faces and Vision*

R.J. Watt
Department of Psychology
and
Centre for Cognitive and Computational Neuroscience
University of Stirling
Stirling, Scotland

INTRODUCTION

Faces are exquisite signaling devices. A person's face transmits a multicomponent message which can be received by another person and decoded to provide information about the transmitter's age, sex, race, identity, emotional state, thoughts, and so on. In this chapter I will consider the relationships between the physical structure of faces and the computational structure of vision in this context of the signaling process. The process of signaling comprises the three different activities of coding, transmission, and decoding. In order to signal information, it has to be given a physical embodiment in a code. The code is then transmitted through some physical medium which connects transmitter with receiver. Finally, the receiver decodes the physical events in the medium and thereby reconstructs the information.

In designing or selecting a code, the most important consideration is the number of states that the code can occupy, since this determines the number of different items, or the amount of information that can be conveyed. A simple switch can occupy two states and can thus be used to code any binary event. A set of n switches can be used to code a signal that has $n + 1$ different states. This is achieved by assigning a number k to each state and then transmitting that state by setting k switches to on and $n-k$ to off. In this simple use of switches it is not necessary to know which switch is which, only how many there are. A set of n

* The work reported in this chapter was carried out with the support of two grants from the SERC Image Interpretation Initiative. The author acknowledges the support of a BP Royal Society of Edinburgh Research Fellowship.

switch states can be used to code 2^n events, if each switch is uniquely identified. The identification can be explicit, where each switch is given a label which is transmitted along with the switch state. The identification can be implicit, where the identity of each switch is only revealed by its context. An example of the latter is a sequence in time or in one-dimensional space. In a list of this type, each item has an ordinal position which is defined by its ordering relations to (e.g., before/after) each of the other items.

There are a number of switches on human faces: hair, forehead, eyebrows, eyes, and mouth being most prominent. Each of these may assume one of a number of different states, although some may be changed more rapidly than others. With these various features human faces have a structure which can be modified in a myriad different ways to provide a large number of potential messages.

Messages are often intended to be private like a telephone conversation, but they may also be public as in radio transmission. In order to decide how to transmit a message, it is necessary to decide what class of receivers it is intended for. If we wish to communicate verbally, a private message is committed to some secure dedicated medium such as a telephone line, whereas a public message is broadcast in some widespread medium. Faces can also be used to transmit private or public messages, but the same broadcast medium of reflected light has to be used for both public and private messages, and so alternative strategies have to be employed to select the appropriate class of receivers. Generally, private messages are transmitted with small facial gestures with the expectation that the receiver will be looking directly at the face and will either be very close to it or will observe the face for a longer than usual length of time. Public messages are broadcast with more obvious and easily detected devices and gestures.

Signaling devices have to ensure that their transmissions are not hopelessly corrupted by environmental factors such as noise, reflections, attenuation, and so on. This is a particularly serious problem for signaling to visual receivers, because the signals are subject to distortions in the mapping from three dimensions onto a two-dimensional image. Furthermore, if ambient lighting is employed, then the actual pattern of the received signal can be radically altered by illumination factors that are not under control.

Next we come to consider the receiver's part in the signaling process. There are some simple aspects to setting communication channels working. The receiver has to be activated by a request to send from the transmitter and has to be tuned or set to the correct channel characteristics. Then the receiver has to acknowledge the request and signal back that it is ready for communication. Obviously, the success or failure of signaling now depends on the coded message being recovered in a suitable form and then decoded or interpreted correctly. In the case of facial messages, the initial request to send is usually in competition with others, and a whole fashion industry has been created to aid individuals in this competition. Once the sender has been noticed and attention has been allocated by the receiver, eye gaze direction and duration are able to signal a readiness for further communication.

The direction and duration of gaze fixations that a person makes have a profound effect on the quality and quantity of visual information that will be received by that person. The eye has a small region of high resolution vision at the center of the visual field, and can only obtain fine scale detail from a small range of directions. The duration of a fixation determines the amount and type of information that can be processed as a spatial pattern. These two factors mean that a person's eyes, and how they move, provide rather direct information about what they are able to see and understand. Fixations therefore provide observable signs of what a person is seeing and can be used as very handy control signals for facial communication.

Faces and vision are an efficient signaling system, and, in this chapter, I will consider how this is the case. In general, I assume that the face was designed with the receiver in mind, and therefore takes advantage of characteristics of human vision that have other functions. To understand the system fully, we shall need to have a description of the types of images that may be formed of faces, considering both the physical structure of faces, with variations from individual to individual and moment to moment, and also the effects of lighting. The second step is to say something about how human visual systems might respond to images of faces. With these two components in place, we can then see how the whole signaling system might work.

EXPOSITION (1ST THEME): IMAGES OF THINGS

We need a few technical terms before proceeding. The eye forms a complete image of the scene encompassed by the field of view. Within this image, there will be the images of a number of things in the scene. The term *image* can thus refer to the projection of the whole scene, or the projection of just one thing within the scene. An image is a two-dimensional function: for every (x,y) position, the image has a value. This value is a light intensity or illuminance. Illuminance is a measure of how much light is reaching a particular place. *Illuminance* of a point in an image depends principally on the luminance of the surface that it came from. *Luminance* is a measure of how much light is leaving a particular place. It is normal, but alas confusing, to talk about image luminance. The luminance of a point on a surface (light leaving) depends on the illuminance of the surface (light arriving) and its reflectance. *Reflectance* is a measure of the proportion of light that a point on a surface reflects. Most lighting can be treated as a mixture of diffuse background illumination and one (or more) point sources of light. The background illumination will affect the overall intensity level in the image, but will have little effect on the form of luminance variation within the image. Any point source of light, on the other hand, contributes to the luminance variations because it is directional. The physics of illumination shows that it varies as the inverse distance of the source, a factor which is unlikely to be significant, and as the cosine of the angle of the direction of lighting, which is a significant factor.

Images of natural or manmade scenes are not random. Instead they are highly constrained in form by the nature of matter and how it interacts with light. Matter is highly cohesive, forming clumps that are bounded by well-defined surfaces. Clumps of matter, that is, things, fill space in constrained ways so that no two things can occupy the same space. Matter is usually connected, which means that the surfaces and the outlines of things are also connected and closed. The particular material that a thing is made from will determine the particular details of its surface: whether it has a highly fractured or textured surface, or whether it is smooth. Surfaces reflect and scatter light to varying degrees, depending on the physical nature of the surface.

These constraints on matter and things cause images to be highly constrained. The image of a single thing is its projection onto a two-dimensional surface. This projection has an outline which is the projection of where the surface folds away from the viewpoint. The outline is a closed curve. The shape of this outline is rather like the perimeter of a cross-section through the thing and thus has a shape that is closely determined by the nature of the surface of the thing. It will be a highly fractured outline if the surface is highly fractured, for example. Within the outline, the image will have luminance variations that correspond to the shape and reflectance of the surface. Texture is due to fixed statistical properties of the surface, and the image has similar fixed statistical properties. The most notable property is the sequential correlations in the luminances. The value of luminance at one point on a surface is highly correlated with the luminance of its neighbors. The luminance variations will be highly variable if the surface is jagged, and will smoothly vary if the surface is smooth.

Because matter is generally opaque, the images of near things will overlay or occlude the images of further things. We thus have a general idea of images as comprising a patchwork of regions with different surface texture characteristics and separated by edges where these characteristics change quite suddenly. Images are thus basically due to two different processes: one acts to create visible edges where things occlude each other; the other acts to fill the images of particular things with patterns of characteristic statistical structure.

We can see the typical structure of such images in more detail by considering how to make synthetic images with closely similar constraints. It is easiest to start with one-dimensional slices through image. We can synthesize these without difficulty by applying two rules. Starting at the left with an arbitrary luminance value we set the next value, rightwards to equal the sum of the preceding value plus a small random perturbation. The next one along is given by the sum of its predecessor plus a new random value. This is the first rule, which I will call the *texture rule*. It corresponds to the process that fills the outlines of the images of things. Every so often we add a rather larger random value, thereby making a sudden large change in luminance. This is the second rule, which I will call the *edge rule*. This rule corresponds to the physical process that creates edges in real images (see Figure 5.1).

As stated, the texture rule produces a Brownian random walk which is a species of fractal pattern. Although the value at each point can be specified by the sum of the value at the previous point and a random variable, the value at that previous point in turn depended on the one before and so on. Another way of describing such a series is to give the expected difference in value between two points, ΔV, as a function of their spatial separation ΔS. If the two points are adjacent, then the difference between them is simply one (Gaussian) random variable, and the standard deviation of the difference is just the standard deviation of the random variable, σ_r. If the two points are separated by two steps, then the second one is equal to the first plus the sum of two (Gaussian) random variables. The difference between them is the sum of two random variables, and since variances add, the standard deviation of this difference is given by:

$$(\sigma_r^2 + \sigma_r^2)^{0.5}$$

More generally, for two points, separated by a distance ΔS, the standard deviation of differences is given by

$$\sigma(\Delta V) = \sigma_r(\Delta S)^{0.5}$$

The values of σ_r and 0.5 are sufficient to characterize the statistics of the texture pattern completely. If we replace the value 0.5 by a variable, H:

$$\sigma(\Delta V) = \sigma_r(\Delta S)^H$$

then we can produce a whole range of statistically different texture patterns. If H is close to one, then the patterns are very smooth; if H is close to zero, then the patterns are very jagged. Figure 5.2 shows a range of such patterns with values of H between 0.9 and 0.1.

This texture rule, with its two parameters of σ_r and H, can be extended easily to two dimensions, although the numerical operations involved are slightly more cumbersome to actually perform (cf. Voss, 1988). In two dimensions, there is no reason why different values of H should not be used for different directions, and this will cause the texture to appear streaked.

We now need to make the edge rule a little more complex. In two dimensions we wish to make the edges into continuous closed curves. If we constrain the edges only to be continuous in the image, then the edge tracks through the image following a random walk. We can describe the contour that it follows, by giving, ΔV, the expected distance (as the crow flies) between two points on the curve as a

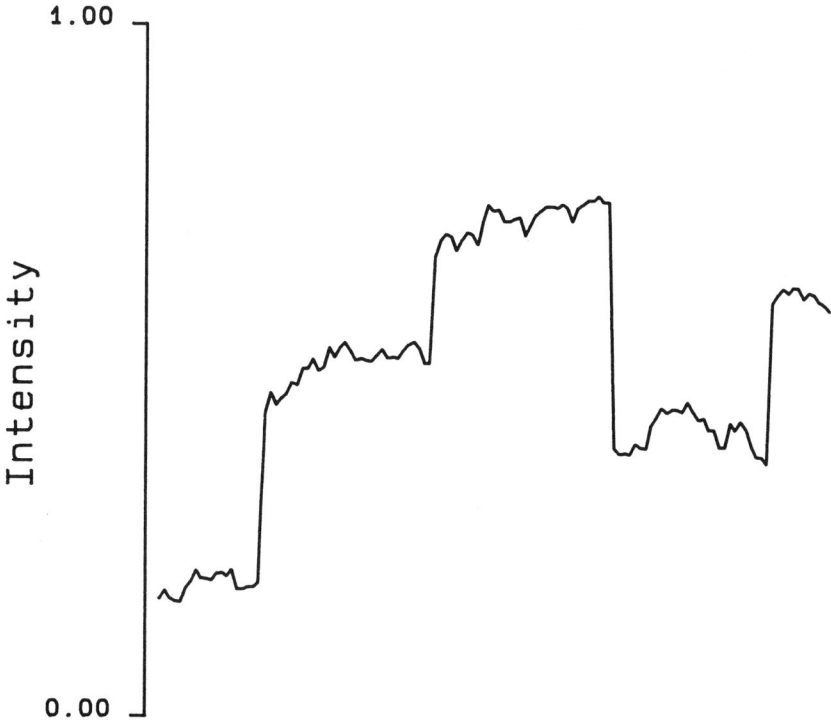

Figure 5.1. A one-dimensional slice through a synthetic, quasinatural image. Notice the smaller variations in intensity that are due to the texture rule, and the much larger discontinuous variations that are due to the edge rule.

function of their separation along the curve, ΔS. For the process that I have described, the form of this is:

$$\sigma(\Delta V) = \sigma_r(\Delta S)^{0.5}$$

which should look familiar. As before we can replace the value of 0.5 by a variable H and then have a parameter which describes how smooth the random walk edge is. Edges like this are not constrained enough because they may not be closed or they may intersect with themselves.

There is a better way of making edges for our synthetic images. If we take a two-dimensional texture pattern of the type described above and with some characteristic H value, and then just take the points where this pattern is (or crosses) zero, then we will have one or more closed, nonintersecting contours as desired. Moreover, the value of H for the contour will be the same as the value of H that was used to make the texture. It would be quite natural to arrange in this way for the outline edge of a patch and its texture contents to have the same value of H. This can be achieved by leaving the texture within the zero-crossing.

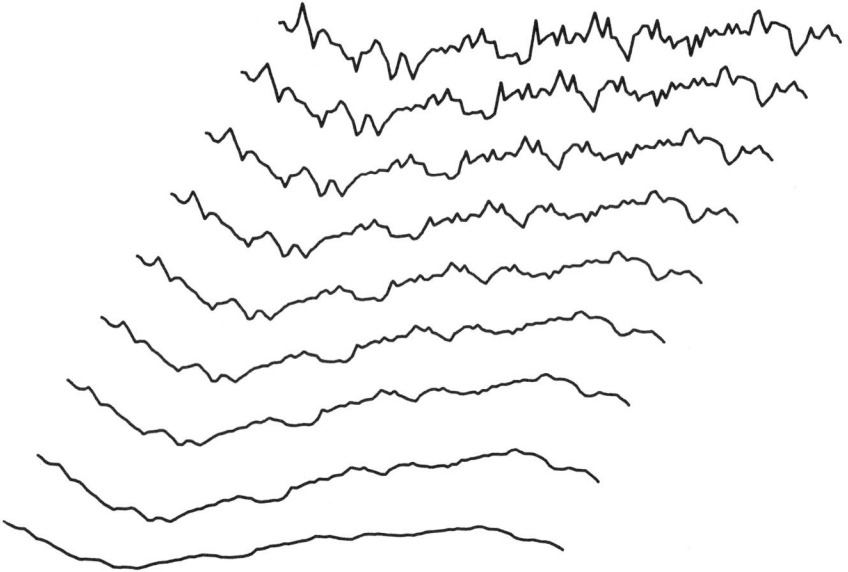

Figure 5.2. This shows a range of Brownian fractal patterns with H values decreasing across and up the figure.

In this section, we have seen how to create synthetic images that have many of the features of natural images. A complete image can be made by creating a patchwork of overlapping texture regions.

EXPOSITION (2ND THEME): IMAGES OF FACES

Consider a spherical head. The image of such a head will be bounded by a circle. If the head is featureless, within this circle the light intensity will vary smoothly. The circumference of the circle will be marked by a discontinuity (sudden change) in intensity between the inside of the circle, which projects from the head, and the outside, which projects from the background. Within the circle, light intensity will vary smoothly in a way which depends upon how the face head is illuminated: on the direction, nature, and number of light sources. There is likely to be at least one point source of light in the environment. This will be in a specific direction from the head and so will make a specific angle of incidence with the surface of the head. This angle varies continuously across the sphere, and so, therefore, does the illumination, according to the cosine law. The intensity will be greatest on those parts of the surface that are face on to the light source and will be least where the light only glances across the surface or where the surface is shaded from the light source. The angle of incidence of light obviously varies between 90° and O°. For a sphere, the variations in angle of incidence are monotonic in all directions away from the place where the angle is 90°. This means that the illuminance is also monotonic away from a single peak, and so the luminance is also constrained to the same pattern because the cosine function is monotonic between O° and 90° (see Figure 5.3).

Of course, a head is not spherical. For a start, it is flattened along the sides and perhaps across the top. The chin juts downwards and perhaps out slightly. This all means that the image of a head is not bounded by the circumference of a circle. The shape of the outline depends on the direction from which the head is viewed. As well as these general departures from being a sphere, the front of a head has a number of very distinctive features: a face. The nose sticks out and

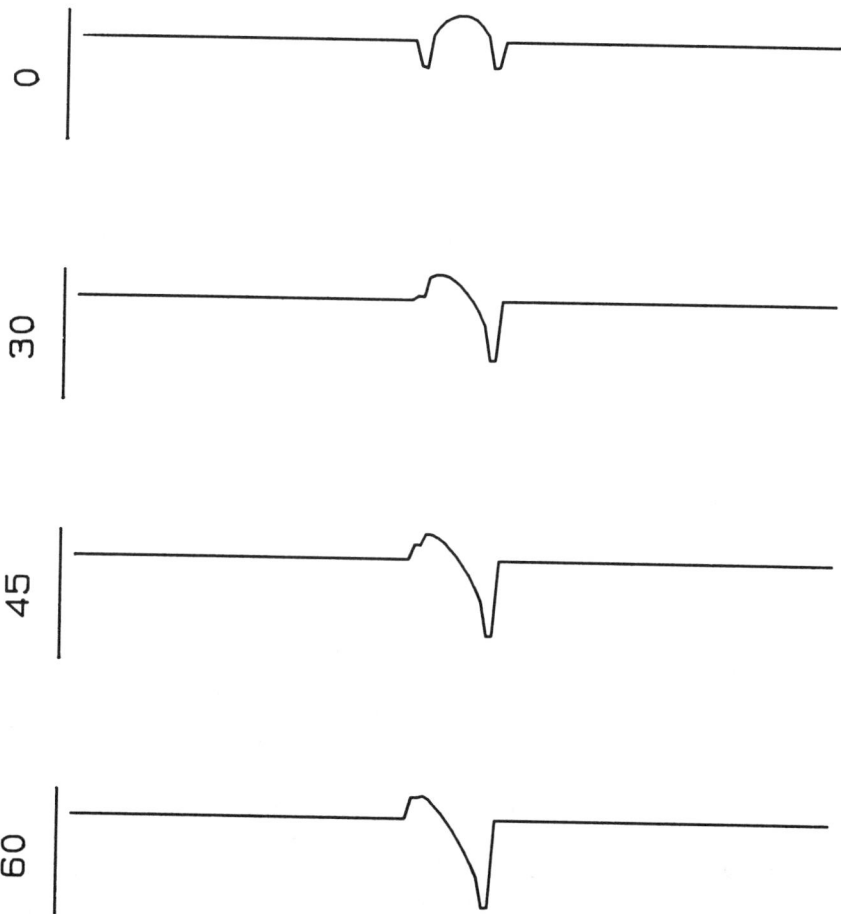

Figure 5.3. A set of horizontal cross sections through images of a spherical head. They differ only in the direction of illumination: The head and viewing direction are unchanged.

cheeks are flattened inwards to where they meet the nose. The eyes are set in deep depressions. The mouth is a slit of highly variable shape, opening to reveal teeth. The ears, attached to each side, are flaps of a highly complex shape. These variations in shape result in some relatively complex variations in the angle of incidence of illumination across the face. Thus, illuminance of the face varies in a way which depends on the shape of the face. The luminance pattern of a face might be said to reflect its shape, although it is more correct to say that the shape reflects its luminance pattern. Wherever the angle of incidence, or, equivalently, the orientation of the surface with respect to the light source, goes through a local minimum or maximum, there is a corresponding local minimum or maximum in luminance.

It is instructive to consider the nose as an example of the effects of surface shape. We can regard the nose as a vertical protruding cylinder with a relatively sharp front and flattened sides. Since it is vertical, illumination direction laterally is relevant. If the source is straight ahead, then the luminance has a peak across the front of the nose which is face on to the light. The sides of the nose have very much lower luminances because they are oriented away from the light source. The cheeks face the light source again and have higher luminance as a result. When the light source is to the side a different pattern obtains. In particular one side of the nose has a high luminance, and the other is likely to be shaded (see Figure 5.4).

If the surface of the face has variations in reflectance, then these also cause variations in the luminance of the image, because luminance is determined by the product of reflectance and illuminance. Where reflectance is low (the surface is dark in color), then luminance will be lower than it would otherwise be. The idealized face can be improved by providing it with some variations in surface reflectance. The hair, eyebrows, eyes, nostrils, and lips all have slightly different reflectances from the basic skin color (see Figure 5.5). The effect of reflectance variations is to modulate the surface luminance. Reflectance features are important, because they will always cause luminance variations, irrespective of lighting, whereas surface orientation only gives rise to luminance variation when the lighting has a strong directional component. The features on a face are mobile: they change shape as the person talks, thinks, breathes, walks along. They are also deliberately manipulated to signal a facial expression: the eyebrows might be raised; an eye might be momentarily closed; the shape of the mouth can be altered.

The image effect of a complex surface such as a face is determined by its shape, reflectance and illumination, but also by the direction from which it is viewed. Viewing direction is irrelevant for a sphere which is the same in all directions, but is obviously very important in determining the layout of the parts of the face image. There is a three-dimensional continuum covering all possible viewing directions. The three dimensions are easiest thought of as corresponding to the three different axes through which a face could be rotated. One lies through the ears, as when nodding; one lies vertically from neck up to scalp, as when

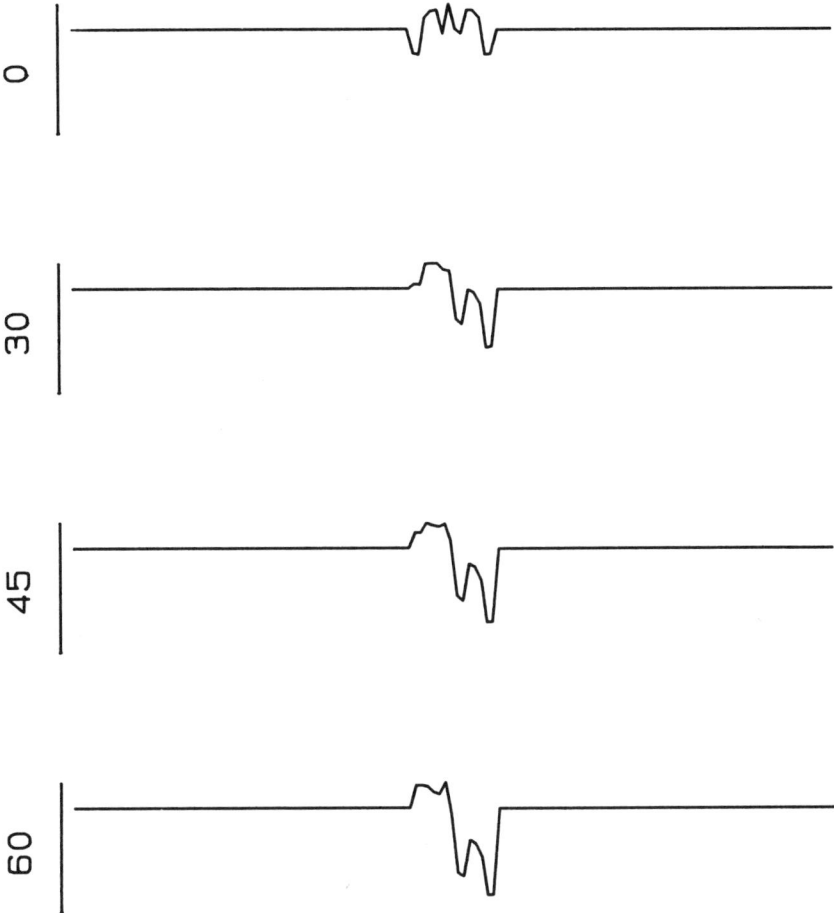

Figure 5.4. A set of horizontal cross-sections through images of a more realistic cheek/nose model of a head. They differ in the direction of illumination.

shaking the head; the third axis lies from the nose backwards as when cocking the head to one side. Any viewing direction can be specified by giving the directions of these three axes with respect to the two axes of the image. Note that there is nothing special about these three; any other set of three that were mutually perpendicular would suffice equally well.

A three-dimensional continuum is a big space to have to consider. It is not really as free as it sounds, however, particularly if the transmitted message is the important object. Under most social situations, meaningful communication only occurs between faces that have nearly the same orientation in two of the three axes: those running through the ears and from the nose backwards. It is rotations about the neck that are most free. This axis lies vertically in a plane about which the face is nearly symmetric, so that even this one dimensional continuum is

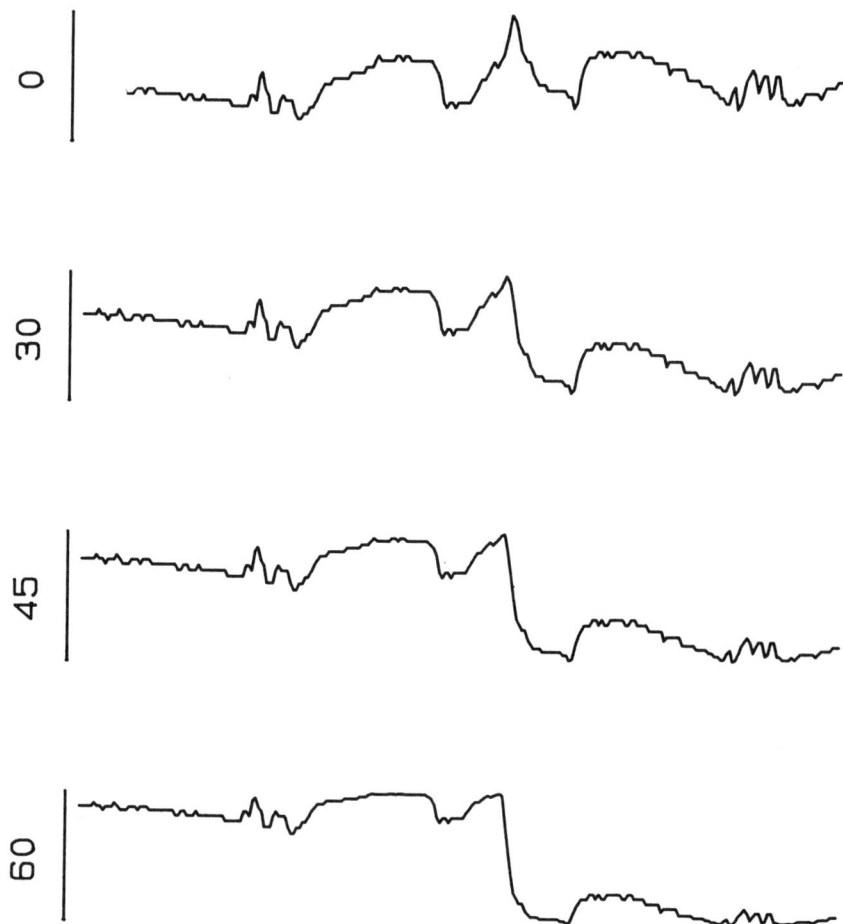

Figure 5.5. This figure shows a set of horizontal cross-sections through images of a front-illuminated face. As in the previous figure, they differ only in the angle of illumination.

degenerate. Moreover, rotations about this axis have less effect than rotations about other axes. Consider, for example, the eye sockets. These are rather like horizontal cylindrical indentations. They are quite deep and are vertically narrow, so that the eye itself becomes effectively occluded by relatively small rotations of the head about an axis through the ears. However, they are horizontally long, and are open at one end, so that very large rotations about the free axis through the neck are needed to equivalently occlude the eye in this direction. The mouth is a similar indentation. The nose, on the other hand, is a protuberance, and, logically, inverse considerations apply. Because it is narrow horizontally but long vertically, it occludes itself under relatively small rotations about an axis

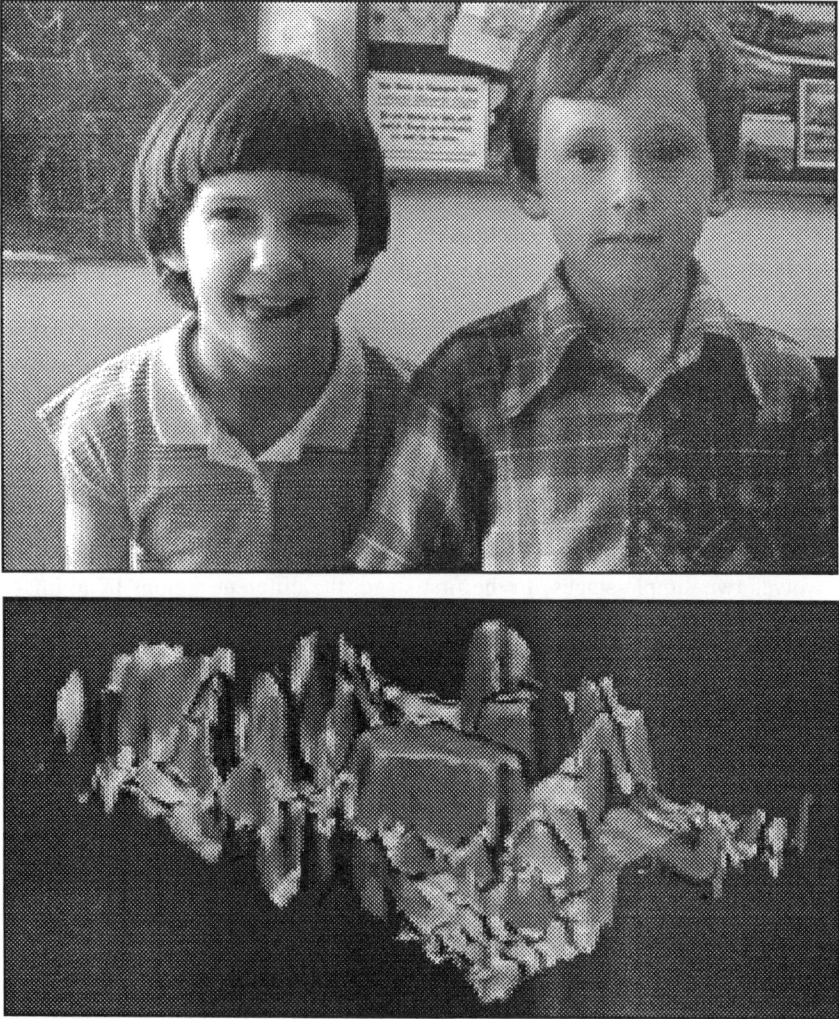

Figure 5.6. This figure shows the intensity profile for a full image of front-illuminated faces plus the image itself. Where the landscape appears high in altitude, the image itself has high intensity.

through the ears, but once again only under extreme rotations about an axis through the neck.

DEVELOPMENT (1ST THEME) LOOKING AT THINGS

The simplest way to describe the activity of the human visual system is to say that it is concerned with the detection, characterization, and interpretation of pattern and structure in the optical image formed on the back of the eye. This sentence can only be said to have meaning once the terms *pattern* and *structure* are defined. It is easier to approach this informally.

A television screen of white noise, where every point is set to a luminance that is random and is independent of all other points, could be said to have a pattern. Within the outline of the display each point is a sample of intensity drawn from a distribution, which itself is stable over the whole screen. This stability lies within an enclosed region which provides structure to the pattern. If, however, the whole visual field were filled with white noise, then there would be no structure to the stability, and it would be meaningless to talk about pattern. Most visual images that we are confronted with contain regions within which there is statistical or pattern stability and between which there are structural outlines. Whenever there is stability in the image, then the whole region can be supposed to have a common physical origin. Thus, much of our power of vision can be construed as determined efforts to identify the structures and the stabilities in an image and to characterize their image properties so that physical properties can then be inferred. I will now describe the simplest ways in which a visual system could obtain an informative description of such images as it might encounter. This involves two simple stages. In the first stage, the different regions in an image have to be distinguished. In the second stage, they have to be characterized and described.

Isolating Regions in Images

We have seen how quasinatural images can be constructed. They comprise a patchwork of different regions within each of which the pattern has some stability. The boundaries between regions are only marked by changes in the underlying statistical processes. The different regions are all distinguished by a number of different factors, such as their mean luminance and H value. Locally, the variance of the image is inversely related to H. The basic techniques for calculating mean luminance and variance of luminance are conceptually very similar. Mean luminance is obtained by summing all the luminance values and dividing by their number (the sum of 1.0 over the same support):

$$L_{mean} = \frac{\int I}{\int 1.0}$$

The variance of luminance is also a summation:

$$L_{var} = \frac{\int (I-L_{mean})^2}{\int 1.0}$$

In each case the equation only makes sense once the area over which the summation should occur has been specified. This is where the problem in vision occurs, because it is rarely possible to know in advance where the regions lie and hence which parts of the image to calculate the sums over.

We can replace each point in the image by its local mean or local variance, as calculated over a relatively small area. The edges of texture regions will be found by looking for large changes in these local statistics. In practice it is best to use a smooth weighting function, such as the normal Gaussian, to define the area over which the summation occurs. This means that near points have a proportionately larger effect on the local statistics (see Figure 5.7).

The precision with which local statistics of an image can be obtained is proportional to the square root of the size of sample set from which they are calculated. This means that there is benefit from using large areas around each point. The cost in doing this is that more and more of the statistics that are computed will be from areas that straddle the edges between different texture regions. This has the effect of blurring out the edges as is illustrated in Figure 5.8.

It is a relatively simple matter to find the places where the local statistics in an image are changing most rapidly. The mathematical operation that is used when looking at changes is called *differentiation* (see Figure 5.9).

For technical reasons involving symmetry, when doing differentiation in two dimensions, it is necessary to differentiate the local statistic images twice horizontally and twice vertically. This operation has the special name of the *Laplacian* and is circularly symmetric, a property that no lower-order differentiation operation possesses. Zero-crossings in a second derivative correspond to places of maximum gradient in the luminance signal. When the degree of smoothing is comparable to the size of the regions, most of the regions in the original image will become zero-bounded distributions of response in the convolved signals. The zero-crossings themselves are likely to be misplaced with respect to the edges, and cannot be relied upon, but the mean or centroid of each distribution is much more stable. This technique of isolating regions by using an appropriate

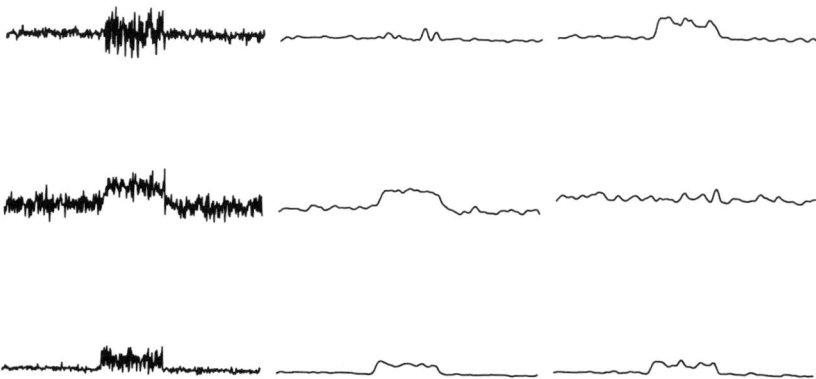

Figure 5.7. This shows some slices through a typical image (left), and the effect of replacing each intensity value by the local mean (centre) or the local variance (right). Notice that the new image is smoother than the one that it replaces.

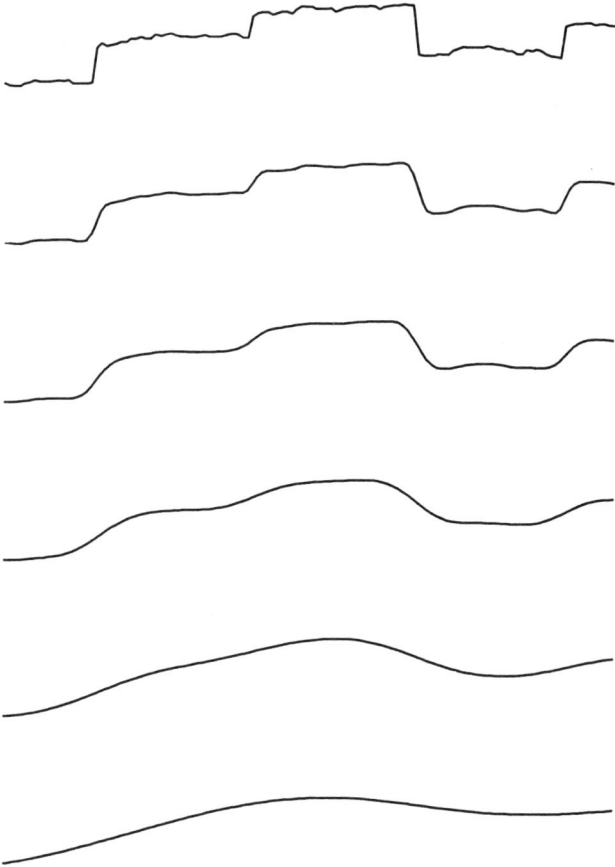

Figure 5.8. This shows the consequence of using Gaussian weighting functions of different widths when calculating local means on the degree of smoothing and the related loss of resolution.

degree of spatial smoothing, or spatial scale, works equally well in two dimensions, but does rely on choosing the spatial scale.

The issue of spatial scale is a complex one. For the present, it is worth noting that Laplacian and Gaussian treated images have a high degree of regularity in their spatial scale structure. If we regard the sign of the output as being the most significant, then the locations of zeroes (zero-crossings) mark where the significant changes lie. It has been shown that the zero-crossings in an image of this sort are nested as spatial scale is varied. This behavior is shown in the space-scale diagram (Witkin, 1983), and an example is shown in Figure 5.11. The figure also shows that the number of sign changes is a reciprocal function of the spatial scale for a fractal texture pattern. The importance of this for visual processing has been considered by Watt (1988). It is thought that edges are

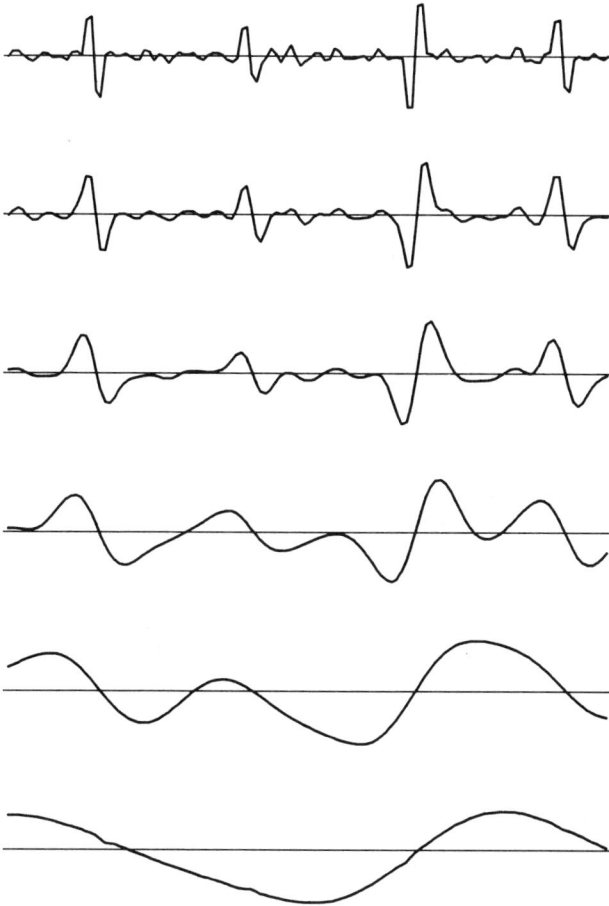

Figure 5.9. This shows twice differentiating the smoothed functions of Figure 5.8. Faint lines mark the level of zero.

characterized by a process that scans through spatial scale from coarse to fine, thereby taking advantage of the nested structure that is guaranteed (Watt, 1987). This means that fine detail emerges after coarser scale structure in the description that is being created.

Characterizing Regions in Images

The outcome of the local statistical analysis of images that I have just advocated is a new type of image where regions are identified as blobs, within which the sign +, or − is always the same. The regions or blobs are separated by zero-crossings, but, as we have seen, these are unreliable, and this must be borne in mind when we come to characterize the regions.

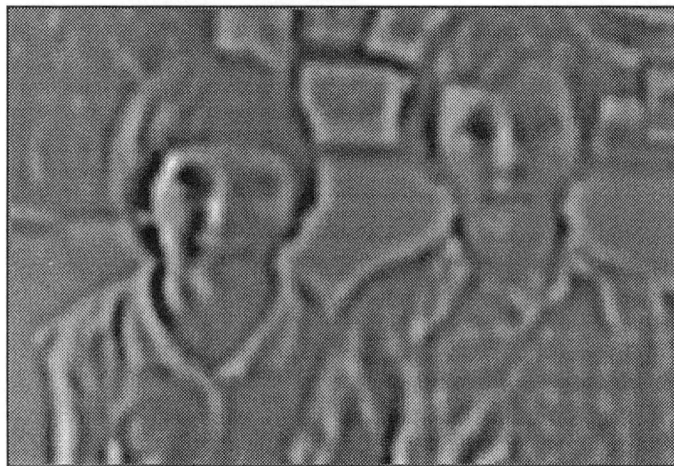

It is convenient to think of the outcome of the characterization stage of analysis as being a set of sentences each of which describes one of the blobs. It does not really matter how we write the sentence down; what does matter is the information that it represents and its precision. There are two types of information to be characterized, which derive from the two rules that were used to generate synthetic images above. The region has certain intrinsic qualities that correspond to the effects of the texture rule, and it has a set of qualities that determine its overall shape, size, and position, corresponding to the effects of the edge rule. These two types of information may not be independent; the *H* value for the outline and for the texture can be the same, for example.

The most reliable descriptors for a distribution like the blobs in these images are those that use all the points in the distribution (Watt & Morgan, 1983). The peak value and its location are each derived from a single point in the distribution and are less reliable than the total mass of the distribution and its centroid or centre of gravity (Watt & Morgan, 1984, 1985). Let us suppose that an image, I', has been created which has only one blob. Then we have:

$$\text{mass} \quad = \quad \int I'$$

$$\text{centroid} = \frac{\int I'.x}{\int I'}, \frac{\int I'.y}{\int I'}$$

The centroid is the (x_c, y_c) location about which the sum of first moments is zero. The first moment of a point (x, y) about a fulcrum (x_f, y_f) is given by the produce of the response value at that point and the distance between point and fulcrum:

$$I'. [(x - x_f)^2 + (y - y_f)^2]$$

If we take just the x-coordinate, the sum of first moments is equal to:

$$\int I'. (x - x_f)$$
$$= \int I'.x - \int I'.x_f$$
$$= \int I'.x - x_f \int I'$$

If we set this sum equal to zero to find the fulcrum point x_c at which the sum equals zero, we have:

$$\int I'.x - x_c \int I' = 0$$

Figure 5.10. This shows the effects of Gaussian smoothing and the Laplacian operation on a typical image. At the top is the original image, beneath this is a Gaussian smoothed version, and at the bottom is the result of then applying the Laplacian operator.

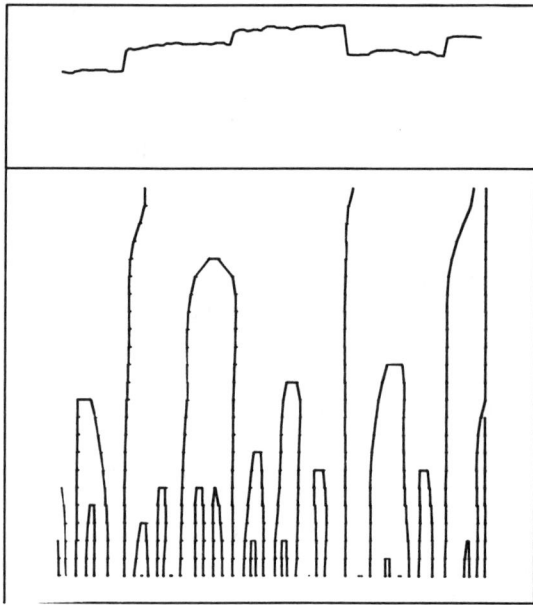

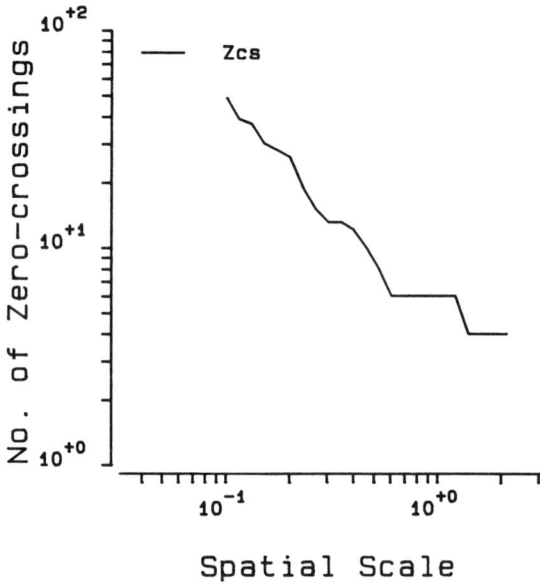

Figure 5.11. The scale-space diagram for the image slice from Figure 5.1 The top panel shows the slice of luminance. The middle panel shows the layout of zero-crossings, at different spatial scales. The vertical axis represents spatial scale with coarse at the top and fine at the bottom; the horizontal axis represents spatial position aligned with the slice above. Each vertical line represents the way in which the locus of a zero-crossing moves as spatial scale is changed. At the bottom is a plot of the number of zerocrossings as a function of spatial scale.

then:

$$x_c = \frac{\int I'.x}{\int I'}$$

Analogous considerations apply to the y-coordinate. By applying an arbitrary rotation of the axes about the centroid, and then proving that the new sum of x first moments still equals zero, this centroid can be shown to be independent of the coordinate system.

There is another way of getting to the same expression for the centroid, starting with the second moments:

$$I'. [(x - x_f)^2, (y - y_f)^2]$$

Taking just the x-coordinates again, we seek the value for x_c at which the sum of second moments reaches a minimum. This is done by differentiating the sum with respect to x_c and equating to zero. So:

$$m_2 (x) = \int I'(x - x_c)^2$$

$$\frac{d}{dx_c} m_2(x) = - 2 \int I'x + 2x_c \int I'$$

equating to zero:

$$- 2 \int I'x + 2x_c \int I' = 0$$

$$x_c = \frac{\int I'x}{\int I'}$$

Analogous considerations can once again be applied to the y-coordinate.

Let us now examine how the sum of second moments about the centroid varies with direction. Changing to polar coordinates, by writing:

$$r \quad = [(x - x_c)^2 + (y - y_c)^2]^{0.5}$$

and:

$$\tan\theta = \frac{(x - x_c)}{(y - y_c)}$$

we have the sum of second moments about the centroid in the direction of Φ:

$$m_2(\Phi) = \int I'(r.\sin(\theta - \Phi))$$

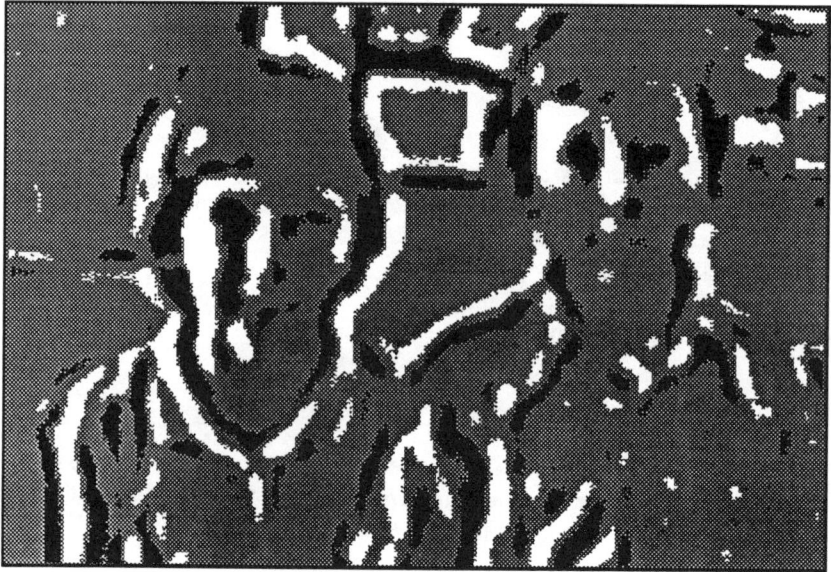

```
DATA
BLOB(    4.00 -339.05     0.68,62.08    40.01,33.15    60.86 )
BLOB(    4.00  227.39   -99.18,13.10    23.27,4.77     89.42 )
BLOB(    4.00  173.35  -160.31,-90.07   23.71,3.39     89.37 )
BLOB(    4.00 -149.08  -104.19,-25.14   18.71,15.45   -70.45 )
BLOB(    4.00  130.24     6.22,74.73    19.64,13.61    -7.20 )
BLOB(    4.00 -102.60   -85.61,16.28    15.77,3.48     89.57 )
BLOB(    4.00  102.52  -108.53,-44.90   15.70,12.81   -64.34 )
BLOB(    4.00  -96.49    24.05,-92.37   18.58,6.90    -82.02 )
BLOB(    4.00   94.53   -21.06,3.73     18.00,17.32    66.92 )
BLOB(    4.00  -92.92  -113.66,33.34    11.50,5.89     78.65 )
BLOB(    4.00   91.20    63.70,68.90    10.20,5.59     80.79 )
BLOB(    4.00   81.29    36.83,-100.71  15.17,5.56     81.17 )
BLOB(    4.00   72.56    -4.09,-85.38   13.62,9.73     73.04 )
BLOB(    4.00   56.65    27.05,-9.21    27.97,10.61    35.02 )
BLOB(    4.00  -56.36  -148.14,-105.82  14.21,6.94    -78.24 )
BLOB(    4.00   56.19   -73.24,16.50     8.01,3.43     80.41 )
BLOB(    4.00   55.97  -122.87,56.43    11.88,6.69     75.91 )
BLOB(    4.00   54.95    91.17,53.57     8.12,2.54    -88.04 )
BLOB(    4.00  -54.26  -170.18,-97.67   19.56,2.80    -87.10 )
BLOB(    4.00   50.34    26.81,110.36    8.99,5.96    -36.32 )
END
```

Figure 5.12. This figure lists sentences describing the larger blobs in the image above.

We can differentiate this with respect to the direction Φ, and equate to zero to find the two directions (necessarily at right angles) in which m_2 is minimum and maximum. These two are called the *principal axes*. The one about which m_2 is a minimum is the direction in which the blob is elongated; it is the orientation of the blob. The two sums themselves are measures of how long and how broad the blob is.

There is another way of thinking about this principal axis. The process I have described is formally equivalent to finding the best fitting straight line for the blob. To find a line

$$y = bx + c$$

which fits a set of points (x_i, y_i) best, we wish to obtain values for b and for c which minimize the sum of the squared deviations, orthogonal to the line. (In this final respect it is different to the familiar linear regression technique.) We can easily go one step further and find a curvature term, by rotating the coordinate axes to align with the principal axes, and then finding values for a and c in:

$$y = ax^2 + c$$

that minimize the sum of squared deviations again. This process could be repeated for ever higher-order line functions, but it is doubtful whether terms greater than the quadratic would be numerically stable (Watt & Andrews, 1982; Watt, 1987).

In this way we can represent each blob by a number of parameters that describe its low-order central moment statistics. Such a representation could be written as:

BLOB (Scale, Mass, X_c, Y_c, SD_{max}, SD_{min}, Orientation)

The whole image is then represented by a set of such sentences, one for each blob. Please bear in mind that a set has no intrinsic structure; all sentences are of equal status; none precede; none succeed. No relationships between sentences exist in a set (see Figure 5.12).

DEVELOPMENT (2ND THEME): LOOKING AT FACES

I now turn to consider how basic considerations about vision can be used to illuminate our understanding of the processes involved in face perception. The general view that emerges is that most information about facial messages is to be found in images of faces with the minimum of processing required for its extraction.

The consideration below is carried out feature by feature, for convenience. If we treat the various features of the face as independent transmitters of messages, we find that many messages are transmitted simultaneously by several different features. There are obvious reasons for why such redundancy is advantageous. In each case, there is some interest in the spatial scale of analysis at which the feature may be recovered. This spatial scale is different for different features.

Gross Appearance

The head has an overall shape that is revealed to a certain extent by its outline. This shape varies from head to head and is thus a clue to the identity of the individual. Some heads tend to be narrow, others wide; some heads have much-enlarged foreheads; other have protuberant jaws. It is thought that several such dimensions tend to covary because of the constraints imposed by the physical processes of bone formation and growth (Enlow, 1982). In addition to shape, the gross appearance of a face depends on hair styles, and these are also revealed in images of heads. Scalp hair is a strong clue to the identity of an individual, being formed into many possible styles. Facial hair is also a highly salient part of the gross appearance of a person's face.

At a very coarse scale of visual analysis, the image of a head is represented by a small number of blobs, and the shape, size, and spatial relationships between these are determined by gross appearance factors. In particular if the hair has a different reflectance from the skin there will be blobs that are directly related to its styling. It is not known how variable the form of the image is at these coarse scales, and hence how variable the representation is both in terms of the actual number of blobs and in terms of their relative parameters. The ratio of the variability of the representation under changes of lighting, facial expression, and viewpoint within one individual compared to its variability across different individuals is a critical piece of information that is currently lacking. The information about gross appearance could be used.

Eyebrows

The eyebrows are highly salient features in the upper half of the face. Unlike those features that contribute to the gross appearance and which are relatively unchanging, the eyebrows are both mobile and deformable. Eyebrows can move up and down and also, to a lesser extent, side to side. They can change in their curvature and in their orientation.

Ekman (1979) has described the basic actions of the eyebrows. In addition to a neutral baseline condition, he distinguishes three elemental eyebrow configurations and a further four formed by combining the elemental ones. These eight configurations are used in a number of different situations, presumably as signals. Two in particular, numbers 4 and 1 + 2, are used in conversation between people either to aid punctuation and emphasis or to signal comprehension or its lacking. Number 4 has negative connotations; 1 + 2 has positive ones. All seven departures from baseline eyebrow configuration are used to signal emotional state, although 2 and 2 + 4 are rare (see Figure 5.13).

If we regard the eyebrows as two elongated markings and seek to describe their various states, then there are three dimensions along which they vary

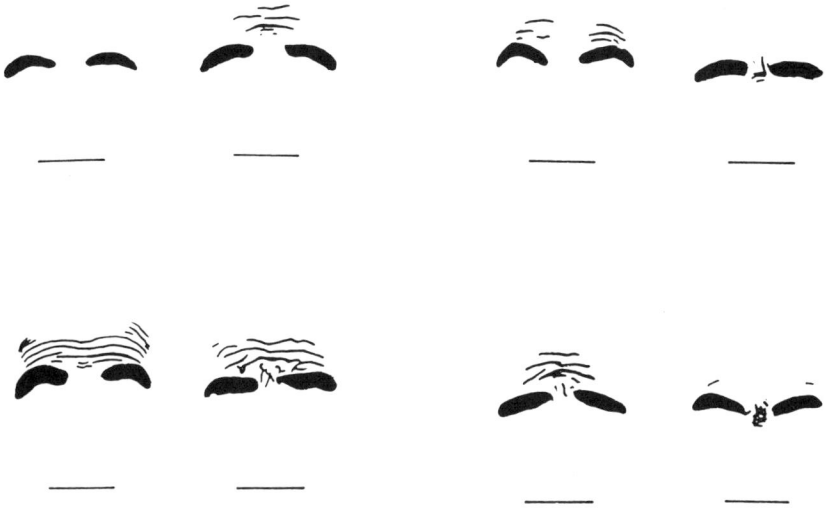

Figure 5.13. This figure has been adapted from Ekman, and shows just the eyebrows as blobs.

systematically. These are planar curvature, orientation, and distance from their corresponding eyes. The two configurations that are used in conversation are on opposite sides of the baseline with respect to their curvatures and distance from the eyes and are thus markedly discriminable. At modest spatial scales, the eyebrows and associated structures give rise to blobs in the image that are remarkably similar in shape to the schematic eyebrows from Ekman (1979). Measurements of the curvatures, orientations, and positions of the blobs that derive from the eyebrows would serve to distinguish quite reliably the various configurations in any one individual. It is not known whether the values obtained would be the same for different individuals or whether the measurements have to be related to the mean for each individual.

Forehead

The forehead is a relatively flat expanse of skin above the eyebrows and below the hair (if any). As the skin loses its elasticity, the forehead responds to eyebrow movements by wrinkling. According to Ekman (1979), the extent and disposition of wrinkles are further indicators of the eyebrow actions. If we add a simple measure of this to the other three image dimensions of eyebrows, then they become further distinguishable. How textured the image of the forehead is will suffice for this measure, combining the two parameters of spatial extent and amplitude of texturing.

Eyes

Beneath the eyebrows are the eyes. These are also highly mobile and obviously provide a strong cue to the direction in which someone is looking. This in turn indicates whether the person over there is talking to you or the person beside you. Closer to, it is thought to be a signal of how a conversation is proceeding, the amount of time spent in eye contact being some measure of the listener's interest and attention.

The eye has a curious pattern of reflectance, with a dark iris and pupil surrounded by a very light sclera. Seen through the eye aperture in the face, the appearance is of two white triangles either side of a dark circular blob. The limbus, or edge contour, between the iris and sclera has a very high contrast. Moreover, as the eye moves the amount of white sclera visible on one side is decreased while the amount visible on the other side is increased. At a sufficiently fine spatial scale of analysis the three parts of the eye will be resolved into three blobs, two light and one dark. There are two cues to the eye direction available: the relative location of the dark blob with respect to, say, the light blobs. Alternatively, the ratio of the masses of the light and dark blobs might also suffice (see Figure 5.14).

A number of investigators have examined, experimentally, how accurately observers can judge the direction of gaze of another person. Gibson and Pick (1963) found that, at a distance of 200cm, subjects could reliably discriminate if a person was looking at them (bridge of the nose) or just to one side of their face (10cm away). This corresponds to a deviation of gaze of 2.8 degrees. Cline (1967) obtained a standard deviation, which is equivalent to a sensory threshold, of 3cm at a distance of 122cm for a full face view and of 5cm for a three-quarters view. The former of these corresponds to a deviation of gaze of 1.4 degrees. If the radius of the eye is 15mm, then a deviation of 2.8 degrees corresponds to a lateral shift of the limbus of 0.72mm, and of 1.4 degrees, 0.35mm. These shifts at viewing distances of 200 and 122cm, respectively, correspond to visual angle changes of 1.2 arc minutes and 1.0 arc minutes respectively. The angular subtense of the visible eye, which measures about 3cm, at 122cm is 1.4 arc degrees, and the lateral deviation sensitivity of 1.0 arc minutes is then equivalent to a Weber fraction of less than 5 percent, which is about as good as could be expected for this type of judgment (cf. Westheimer & McKee, 1977; Watt, 1984).

Mouth

The mouth is a very flexible feature of the face. In addition to its physiological functions, it can also serve as a signalling device for transmitting information both about emotional state and about speech. The latter has been given some analysis (cf. Campbell & Dodd, 1987) and a summary of some of the work is given by Brooke in this volume.

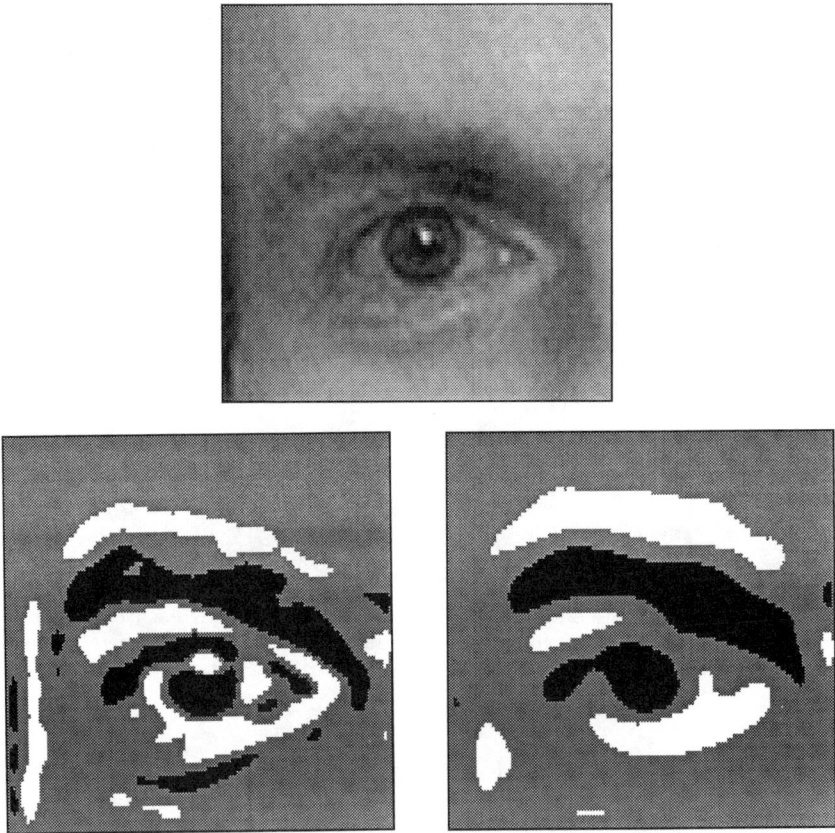

Figure 5.14. This figure shows an extract from a face image containing an eye. Beneath the original image there are two processed versions showing the blobs discovered at two different spatial scales.

As the mouth opens and closes to emit and modulate sound, the appearance of the mouth and lips changes, providing potential visual cues to what is being said. Some of these cues have been measured from various different recording systems, including video cameras, X-rays, and even stone castings. It is known with reasonable certainty that the width of the mouth opening varies between about 15cm and about 45cm, adopting these two extremes or a midway width of about 25cm for most vowel sounds (Fromkin, 1964). The height of the mouth opening varies from 5cm to about 20cm for the same vowel sounds. There is some degree of correlation between width and height, which means that their product, which is proportional to the area of the opening, does not usually confound narrow high openings with wide thin openings. The range of areas of mouth openings for the different vowel sounds has been reported to be around 20:1 (Montgomery &

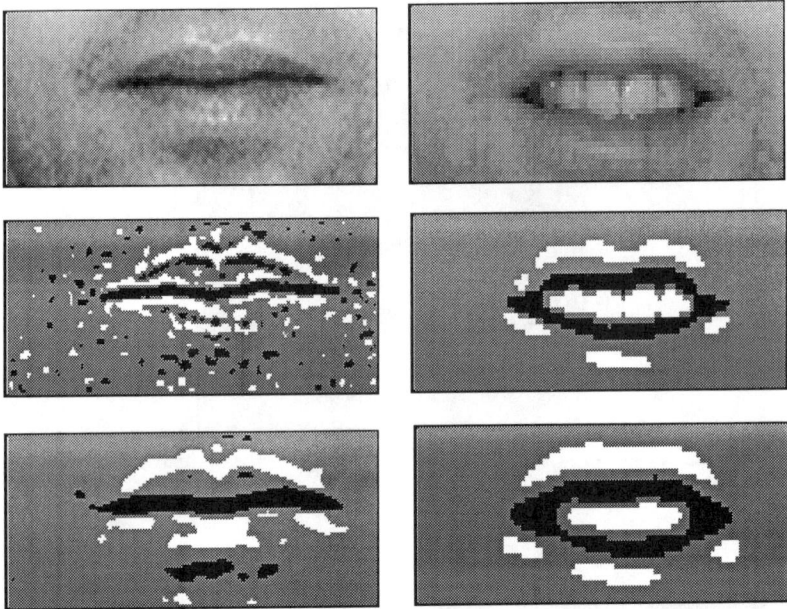

Figure 5.15. This figure shows two extracts from face images each containing a mouth but with different expressions. Beneath the original images there are two processed versions showing the blobs discovered at two different spatial scales.

Jackson, 1983). The study of Montgomery and Jackson (1983) noted that there are no lip configurations that invariably cue specific vowels, although an "u" sound is generally associated with a small mouth opening. These two variables of mouth width and height are found to be weak predictors of how well people can lipread, indicating other cues must be present and used. McGrath, Summerfield, and Brooke (1984) made a detailed study of the cues available and concluded that two extra sources of information should be considered. These are the visibility of the teeth and perhaps tongue, and the shape of the lips. It seems likely that different speakers have different mouth shapes to start with, and that it is chang- ing mouth shapes that are important rather than the static shapes that photographs capture (see Figure 5.15).

When the lips open, they reveal the much darker inside of the mouth. Even at a relatively coarse spatial scale this will give rise to a dark blob. The two standard deviations, long and short, of the blob are a measure related to the size of the mouth opening, and the curvature of the blob is a measure of whether the corners of the mouth are turned up or down. Changes in these parameters over time might well be powerful cues for lipreading. If the mouth opening reveals the teeth, then these will also be manifest as a discrete blob, even at relatively coarse scales.

RECAPITULATION

The general theme of this chapter has been the communication of messages using faces and vision as transmitter and receiver. It has proved necessary to use only a minimal understanding of vision as a pattern describing device, and a minimal characterization of the physical structure of faces. From these modest starting points, it has proved possible to study connections between the transmitter and receiver. In this section I shall summarize the information that I have developed above, but in a different arrangement.

Most social interactions between people have two natural progressions that are mildly correlated. People tend to come closer together in space during the course of an interaction and the messages tend to become more personally detailed and significant. The extent of and rate of approach depends on the social situation and relationship and is obviously controlled by exchange of message. Several of the messages are facial, and this provides a vehicle for a summary of the relationship between faces and vision. What aspects of faces, and which messages, are visible at different distances? I have chosen a range of distances and asked what can be seen in a face at each distance. The answers that I give are, for the most part, pure speculation based on known visual abilities such as two-point resolution. As such the answers are probably to be taken as upper limits on performance, and overlook, for example, the potentially significant context cues that are also available.

The basic effect of viewing distance is a change in the size of the image. Given the optical performance of the eyes, which set limits on the detail that can be imaged onto the retina, changing image size is equivalent to blurring the image without changing its size. For illustrative purposes, and to avoid having to view the figures in this book from inordinate distances, I have prepared some versions that show the effects of a range of distances on the information available, and all from the comfort of your armchair. These have been obtained by isotropic Gaussian blurring. In order to present the resultant images with fairly natural sharp edges, the grey levels have then been quantized to half standard deviation intervals about the mean grey level. This technique relies on the background having a luminance roughly the same as the mean luminance of the face. The quoted equivalent distances are independent of both your viewing distance and the image width.

In analyzing the effects of image size, three spatial scales are particularly significant.

1. The resolution limit is equivalent to a spatial scale with a standard deviation of 0.7 arc minutes (Watt & Morgan, 1983). In the sections that follow, it is this scale that is used to define what information is present and therefore might be used, although this must be qualified by the next two points.

2. Under prolonged inspection, the most significant spatial scale, with the greatest contrast sensitivity and greatest contribution to spatial localization, has a standard deviation of 2.8 arc minutes (Watt & Morgan, 1985), which is a factor of four coarser than the resolution.

3. Watt (1987) has described evidence which would seem to indicate that, for exposure durations less than 500ms, the most significant spatial scale is larger still. If most eye fixations last for around 100–200 ms, the predominant spatial scale is four times larger still at around 11 arc minutes. The implication of this finding is that being able to examine a face for a longer than usual time leads to a finer spatial scale analysis which in turn is equivalent, in some respects, to being nearer to that face.

Watt and Morgan (1985) describe how the range of spatial scales between the predominant scale and the finest scale are combined and used. Watt (1987) has suggested that the predominant spatial scale obtains a fully geometric representation, whereas finer scales tend to receive statistical analysis and representation. In the light of this, the distances quoted below are not to be taken too seriously; it is probable that much nearer distances would be required to reveal the same information in real social situations. The study of facial communication would be an ideal means of examining the dynamic hypothesis of Watt (1987).

Similar pictures were given to a number of colleagues here in the psychology and computing science departments who were asked to say what they could about the people depicted. They were shown in sequence starting with the most blurred. The remarks obtained tended to become too personal and embarrassing for equivalent viewing distances of 2.5m and closer. In the discussions that follow these informal remarks were used to suggest what might be visible at different distances. The suggestions so obtained are always clearly identified as such.

1. 160m. The typical head has dimensions of about 280 * 160mm. From a distance of 160m, these figures give visual angle dimensions of 6.0′ * 3.4′. From such a distance, gross appearance of the face is visible, as can be seen from the figure, in which a number of different faces have been blurred as described above. Very little else is apparent. Some indication of gender might be gleaned from the shape and from the distribution of hair. Subjects at Stirling were hesitantly correct at identifying gender at this distance (see Figure 5.16).

2. 80m. From a distance of 80m, the head subtends a visual angle of 13′ × 7′. At this distance, the symmetry of the image would give an indication of the angle of viewing. By now, the overall shape of the head is available from the dispersion of blobs. In addition, there is the beginning of the appearance of blobs that are derived from structural features of the face itself such as eye sockets and cheek bones. In the informal experiments here, subjects were all able to identify the individuals depicted, each of whom was distinctive in appearance.

It is important to realize that, at this distance, it is only a very sparse represen-

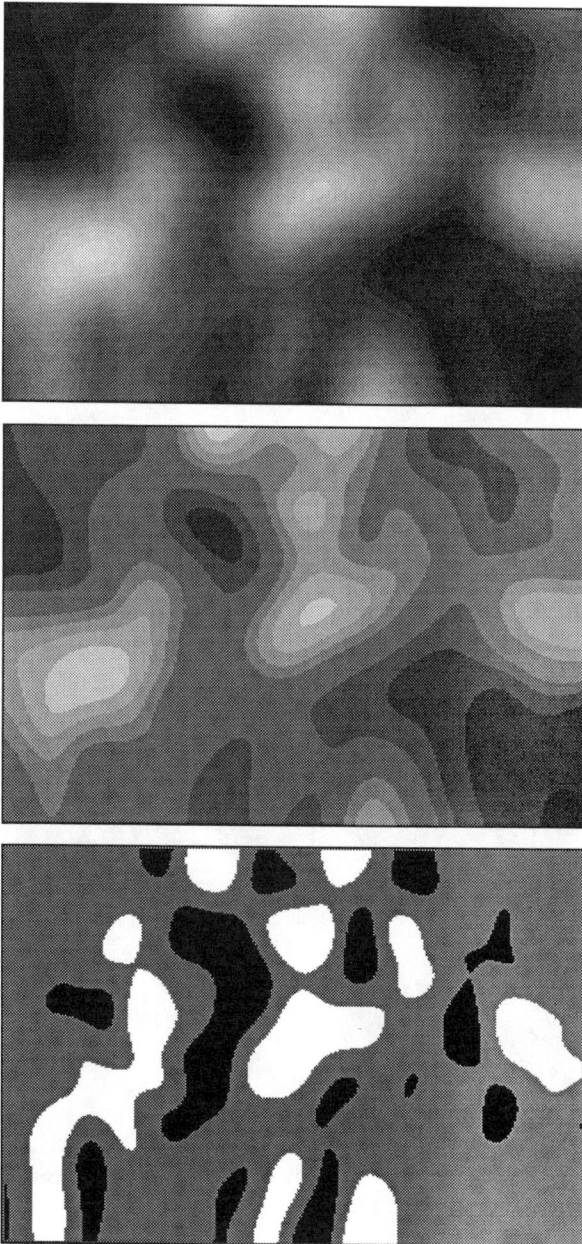

. **Figure 5.16.** A representation of the information visible from a distance of 160m.

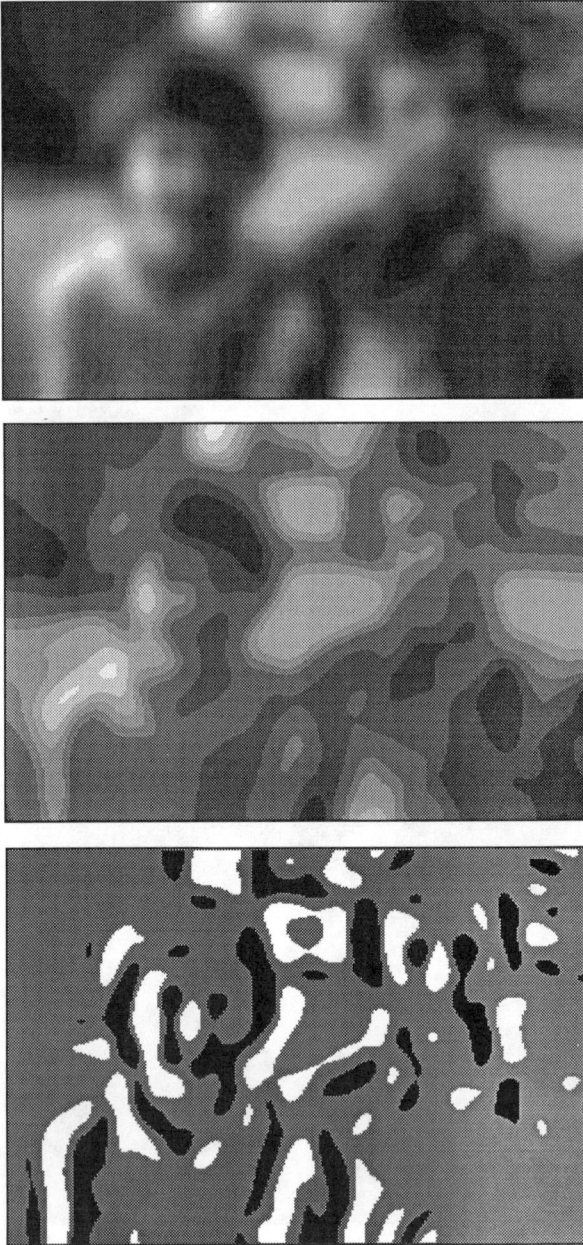

Figure 5.17. A representation of the information visible from a distance of 80m.

Figure 5.18. A representation of the information visible from a distance of 40m.

Figure 5.19. A representation of the information visible from a distance of 20m.

Figure 5.20. A representation of the information visible from a distance of 10m.

tation that is available to support the identifications that these subjects were making. At this distance a description of the face is still not possible. It therefore follows that only relatively distinctive and highly familiar faces could be identified (see Figure 5.17).

3. 40m. From a distance of 40m, it is possible to identify faces with a high degree of accuracy. Hayes, Morrone, and Burr (1986) found that observers could reliably and correctly relate different views of faces with information available at this distance, even though the faces were chosen not to have distinctively different hair styles or overall shape.

Informal experiments here suggest that observers might be able to discriminate whether the mouth is open or closed from this distance. The mouth cavity height ranges from 0 to 1.3 arc min (=15mm) during speech, with the different vowels falling in the range 0.4 arc min. (=5mm) to 1.3 arc min (data from Fromkin, 1964). The illustrations of facial expression in Ekman and Friesen (1975) show mouth openings that have a height of 20mm at most. From 40m this would correspond to 1.7 arc min., and could be visible. If the teeth are revealed, then these will give rise to a blob of their own in the image response (see Figure 5.18).

4. 20m. From a distance of 20m a great deal of information is visible in a face. The eyebrows are resolvable from this distance. They have a size of about 9 arc min by 1.7 arc min (=50 × 10mm), and this is long enough for their curvature to be readily discriminable. The mouth has a normal resting width of around 50mm, corresponding to about 9 arc min. Its curvature will also be readily discriminable. The lips are much less obviously visible, depending on lighting.

The eyes are just about visible. The iris has a diameter of about 10 mm, and the eye socket has a width of about 25mm. These distances correspond to visual angles of 1.7 and 4.3 arc min, respectively. A minimum interperson distance of 600mm would correspond to a rotation of the eye through an angle of 1.7 degrees. This corresponds to a lateral shift of 6mm in the eye limbus; 6mm in turn corresponds to a visual angle in the observer of 12 arc seconds. This deviation would be on the limits of discriminability. From 20m it might just be possible to tell if someone was looking at you or your neighbour (see Figure 5.19).

5. 10m. From a distance of 10m, the head subtends a visual angle of around 1 arc degree and therefore fills the fovea. At 10m, the lips are clearly visible and lipreading should be possible, even for relatively indistinct movements. Facial wrinkles with a typical period of 5mm will be resolved. There is very little signal available in a face that cannot be seen from this distance, if the face is fixated in the center of gaze for a few seconds (see Figure 5.20).

CODA

We have considered faces and vision as a signaling mechanism. A face can signal who a person is at various levels, such as race, gender, age, and individual identity. Its parts can be used to signal attention and control conversation, to signal emotional state, and to aid in speech transmission. In order to understand how these might be achieved, we have only needed to make a few simple assumptions about the type of visual representation that is available. With these assumptions, which can be supported by psychophysical evidence, and with a few known constraints on the performance limitations of human vision, it has proved possible to speculate about the distances from which these different signals are visible. A natural order is manifest in the sequence. As you approach someone, the following sequence in time is predicted:

gross shape, gender
angle of viewing
identity, if it is someone familiar
identity, more generally
mouth opening, gross expression
eyebrows, emotional expression
direction of gaze, lipreading cues

I will finish this chapter by discussing how the information about different facial features can be derived from a visual description of the image. Let us suppose as a starting point, that we have a complete set of blob descriptions at each of a range of different spatial scales. Recall also, from the introduction, the fundamental difference between a message which is a set of switch states and one that is a list of switch states. Each blob is basically just a multistate switch, and since they are all just recorded as blobs, there is nothing to uniquely identify each, at least in explicit terms. Recall, finally, that we noted in the introduction a way in which items in a set could be implicitly transmitted as a list by using some form of sequence. Stated rather baldly, the vision problem for faces can be construed as the problem of knowing which blob or blobs to use for different purposes.

Each blob has a number of parameters that define its location in some high-dimensional space, just as two parameters would locate it in a two-dimensional space. It might be feasible to maintain some record of where each blob that is used for each specific function should lie and within what limits. This would only work if the blobs occupied the space very sparsely and were always located within the same region of the multidimensional space. Neither proposition seems particularly likely, and it is probable that some form of absolute ordering by location will not suffice, even when restricted to only full face views.

An alternative, that requires and deserves some examination, would be to use the spatial scale of analysis to define a hierarchy. Of the coarsest scale, where there are very few blobs, their spatial layout is constrained to fall into one of a very small number of configurations. At a somewhat finer spatial scale, each blob can be assigned a parent at the coarser scale and have its location in multidimensional space represented with respect to that parent. For any one configuration at the coarser scale, there is in turn only a small number of ways in which the finer blobs can be laid out. The process can obviously be repeated, representing the information in each blob at each spatial scale always with respect to a parent blob at the next coarser scale. The resultant structure is a tree graph or hierarchy which, when examined from a coarse to fine scale, can be regarded as progressively refining the overall representation (cf. Watt, 1988). At the same time, it can also be viewed as narrowing down the possible locations of a target blob coding the information concerning some specific message.

Of course, this is speculation, requiring both computational and psychological research before it can be regarded as a potential model for the relationship between faces and vision.

REFERENCES

Campbell, R., & Dodd, B. (1987). *Hearing by eye.* London: Erlbaum.

Cline, M.G. (1967). The perception of where a person is looking. *American Journal of Psychology, 80,* 41–50.

Ekman, P. (1979). About brows: Emotional and conversational signals. In M. von Cranach, K. Foppa, W. Lepenies, & D. Ploog (Eds), *Human ethology.* Cambridge, England: Cambridge University Press.

Ekman, P., & Friesen, W.V. (1975). *Unmasking the face.* Englewood Cliffs, NJ: Prentice-Hall.

Enlow, D.H. (1982). *Handbook of facial growth.* Philadelphia, PA: W.B. Saunders.

Fromkin, V. (1964). Lip positions in American English vowels. *Language and Speech, 1,* 215–225.

Gibson, J.J., & Pick, A.D. (1963). Perception of another person's looking behaviour. *American Journal of Psychology, 76,* 386–394.

Hayes, A., Morrone, M.C., & Burr, D.C. (1986). Recognition of positive and negative bandpass-filtered images. *Perception, 15,* 595–602.

McGrath, M., Summerfield, A.Q., & Brooke, N.M. (1984). Roles of lips and teeth in lipreading vowels. *Proceedings of the Institute of Acoustics, 6,* 401–408.

Montgomery, A.A., & Jackson, P.L. (1983). Physical characteristics of the lips underlying vowel lipreading performance. *Journal of the Acoustical Society of America, 73,* 2134–2144.

Voss, R.F. (1988). Fractals in nature: From characterization to simulation. In H.O. Peitgen & D. Saupe (Eds.), *The science of fractal images.* New York: Springer-Verlag.

Watt, R.J. (1984). Towards a general theory of the visual acuities for shape and spatial arrangement. *Vision Research, 24,* 1377–1386.

Watt, R.J. (1987). Scanning from coarse to fine spatial scales in the human visual system after the onset of a stimulus. *Journal of the Optical Society of America, 4A,* 2006–2021.

Watt, R.J. (1988). *Visual processing.* London: Erlbaum.

Watt, R.J., & Andrews, D.P. (1982). Contour curvature analysis: hyperacuities in the discrimination of detailed shape. *Vision Research, 22,* 449–460.

Watt, R.J., & Morgan, M.J. (1983). The assessment of visual location: theory and evidence. *Vision Research, 23,* 97–109.

Watt, R.J., & Morgan, M.J. (1984). Spatial filters and the localization of luminance changes in human vision. *Vision Research, 24,* 1387–1397.

Watt, R.J., & Morgan, M.J. (1985). A theory of the primitive spatial code in human vision. *Vision Research, 25,* 1661–1674.

Westheimer, G., & McKee, S.P. (1977). Spatial configurations for visual hyperacuity. *Vision Research, 17,* 941–947.

Witkin, A. (1983). Scale-space filtering. *Proceedings of the Eighth International Conference on Artificial Intelligence, 2,* 1019–1022.

II
Facial Reconstruction and Animation

6

Face Recognition and Recall Using Computer-Interactive Methods with Eye Witnesses*

John Shepherd
Department of Psychology
University of Aberdeen
Aberdeen, Scotland

Hadyn Ellis
Department of Psychology
University of Wales College of Cardiff
Cardiff, Wales

INTRODUCTION

Eyewitness evidence plays an important role in the investigative and in the probative aspects of police work. In particular, a good description of a suspect can facilitate the course of an investigation. Traditionally, where the suspect is unknown to the victim or witness, there have been three methods of obtaining such information. The witnesses can provide a verbal account of the suspect's appearance; they can be shown photographs, or, if a suspect is in custody, an identification parade to see whether they can recognize the suspect; or they can produce a pictorial impression in the form of an artist's drawing made from their description, or a composite generated from using one of the commercial composite systems such as PhotoFIT or Identikit.

---

* The development of FRAME, FACES, and EFIT was supported by contracts between the Home Office and the Department of Psychology, University of Aberdeen. The views expressed above are those of the authors, and may not be those of the Home Office. Work using Mac-a-mug was supported in part by an ESRC Programme Award: XC1525000 to H.D. Ellis. Jean Shepherd produced Figure 6.3, and Diane Ellis Figures 6.4 and 6.5.

Obtaining an adequate verbal description will be a necessary procedure no matter what other aids are used, and the quality of these descriptions can be enhanced by such psychological techniques as the *cognitive interview* (Geiselman, Fisher, MacKinnon, & Holland, 1986). So far little use has been made of computers at this stage of investigations. However, with the advent of image-grabbing facilities as peripherals to desk-top computers, it would be possible for witnesses to use video images to recreate the scene of the crime, which is an integral aspect of the cognitive interview.

In the other two areas of witness memory—recognition of the suspect from photographs, and recall of the suspect using composites—there have been substantial developments in computer applications.

PHOTOGRAPHIC RECOGNITION

A witness can only recognize a suspect's photograph if there is one in the possession of the police, which is likely to be the case if the suspect has a previous conviction. English police forces retain photographs of all people convicted of an offence (in Scotland the forces are more selective), and the prints may be assembled in albums for the use of witnesses. The number of photographs held by a force will run into thousands, or even into hundreds of thousands, depending on the size of the force, and a selection of these will be assembled into albums, where the number may range from a few thousand up to 40,000. For any one witness a special album may be assembled according to the type of crime, and the sex or some special category of the suspect, and the witness may be asked to look at 800–1,000 photographs in a single search, and sometimes more.

Clearly, these procedures give rise to a number of problems. There is the police manpower involved in selecting photographs and assembling them into albums, or in weeding and updating more permanent albums. From the witness's point of view, the task of searching through such a large set of faces may require an effort of attention which may be difficult to sustain throughout the search. Although there are very few studies looking at this problem, there is sufficient evidence from laboratory investigations to suggest that, as the size of the set in which the target is embedded increases beyond a hundred or so, there is an increase in the number of misses and false alarms, indicating some change due either to fatigue, interference, or shift in criterion (Davies, Shepherd, & Ellis, 1979; Ellis, Shepherd, Flin, Shepherd, & Davies, 1989; Laughery, Alexander, & Lane, 1971; Lenorovitz & Laughery, 1984).

Both the demands on police personnel and on witnesses can be diminished by the judicious use of a computerized system. The witness problem could be eased if the number of mug shots to be inspected could be substantially reduced, but in doing so the risk of excluding the suspect would have to be minimized. The demands made upon police time by such a procedure could be alleviated if a rapid means of producing reduced "albums" could be devised.

An early approach to the "pruning" problem was reported by Goldstein, Harmon and Lesk (1972), who developed a database for 255 faces each of which was scaled on 21 attributes by a group of trained raters. Subjects described target faces using the same set of scales, and a search algorithm was used which ranked faces in the data base in the degree to which they matched the description of the target on each attribute. Goldstein et al. tested the algorithm by asking human subjects to describe a target face which was displayed in front of them on the scales on which they had been coded. In the subsequent searches of the database, the target was retrieved in the top 4 percent of ranks in 93 percent of trials, and in first place in 67 percent of trials. An interesting aspect of this work was the provision of an automatic feature selection. This selected, for the subject's assessment, the most discriminating attribute defined in terms of the distribution of values for the appropriate subset of faces. In this way it was argued that the computer-interactive system would combine the ability of the subject to identify the most extreme features in the target, with the computer's ability to select which attributes would be most discriminative. The results indicated that the method resulted in very little improvement in identifying the target. Used on its own, without subjects selecting attributes for description, the technique led to poorer retrieval performance than either the procedure based on subjects choosing their own attributes, or a mixture of the two procedures.

More recently, Lenorovitz and Laughery (1984) have described CAPSAR (computer-aided photographic search and retrieval), a system developed at the University of Houston in which the images in the database are Identikit composites. Identikit is a facial composite system comprising line drawings of single facial features on acetate foils which can be assembled into a whole face by overlaying the foils. The database was produced by selecting 25 examples of chin, mouth, nose, eyes, and eyebrows, and 50 examples of hair, which were then used to generate 335 facial images to serve as a database and targets for the system.

Features were precoded by judges who ranked them along 18 selected attributes, with the mean ranking across 40 judges being used as the database value. In addition, a subset of the features ranked on each attribute were selected as prototypes for display as a graphic scale to subjects during the search phase.

For a search trial, the subject was shown a target composite face. After obtaining free verbal recall of the composite from the subject, the experimenter then selected, from the 18 attributes, any which corresponded to a descriptor provided by the subject. The prototypes for that attribute were shown to the subject, who was asked to indicate the position of the target face as well as the possible range beyond which the target was unlikely to fall. This information was then used by CAPSAR to eliminate from the search all faces whose values on this attribute fell outside the indicated range.

Following this stage, a face was randomly selected from those remaining after the initial pruning, and the subject was asked to indicate, if it was not the target, how it differed from the target. As a result, further pruning of the search set occurred, and a further face was selected from the remaining set. This continued until the target appeared on screen, or the search set was reduced to 50, when the subject carried out a linear, conventional mug-shot search of those remaining.

Lenorovitz and Laughery report hit rates of 53 percent and false alarm rates of zero for CAPSAR, compared with rates of 32 percent and 25 percent for the equivalent measures using a conventional mug-shot album procedure.

Each of the two approaches described so far relied upon "syntactic" coding of features for their database. That is, judges were required to make assessments of a value or a verbal descriptor to encode the attributes of the face for storage. Access to the store from the witness's description also depended upon coding the description the witness provided on the same descriptors.

One disadvantage of such a system is that coding large numbers of faces would be expensive in time, and be prone to unreliability. Ideally, a method of coding faces by physical measurement might overcome these problems, since in principle it would be possible eventually to code faces automatically with such a system.

The Houston research team report one such attempt (Batten & Rhodes, 1978). Their aim was to devise a system which would accept input from an artist's sketch or an Identikit composite for comparison with a set of records in the database. Ten measurements, nine linear and one angular, were adopted on the basis that they represent meaningful dimensions for defining properties of the face, and that they could be measured with satisfactory precision and related to features which were stable, such as interocular distance or the distance from chin to eyes.

For search purposes, measurements from sketches or Identikit composites which had been generated by witnesses were converted to ratios, and a series of transformations were carried out before searching the database of similar measures taken from photographs. A goodness-of-fit measure was obtained between the input and the corresponding set of parameters for the photographs and a ranking of fit was produced. Using a data set of 67 targets, and measurements derived from sketches and Identikits on 62 of these, the mean rankings of the target photographs for searches from sketches was 26, and from Identikits 33. The goodness-of-fit of the sketches and Identikits was also evaluated using direct ratings obtained from a panel of judges. Correlations between these ratings and the rankings based on physical measurements were all low ($-.097$ to $.065$) and nonsignificant (Laughery, Duval, & Fowler, 1977).

From these results it appears that using a graphic input in the form of a sketch or composite, or even of a photograph to search a database of physical measurements, is not very effective. One possible explanation for this result is that the images for input were of poor quality. However, some of the sketches were made by artists live from the target person and were given a goodness- of-likeness rating of about 2.5 on a 6-point scale where 1 was the best likeness. Another reason for the disappointing performance of the system may be that the measures used are not the most appropriate for discriminating among faces, especially when the image for input has low precision.

The approaches described so far illustrate the issues which need to be addressed in developing a mug-shot retrieval system. First the images have to be coded in some way for storing in the database. Laughery, Rhodes, and Batten (1981) distinguish between geometric coding and syntactic coding. Geometric coding entails taking measurements from images either manually or automatically. Laughery et al. (1977), as mentioned above, tested such a system, without much success. One of the difficulties underlying this method is the selection of which of the many possible measurements to use. The criteria of the measurements being constant for a face over time, and being "representative" of the dimensions of a face, appear to be sensible. However, the purpose of the mug-shot databases is to allow a search to be made on the basis of a witness' description. A consideration of the attributes which most often occur in a witness' description of a face shows that the most commonly mentioned features tend to relate to the hair, such as its length and color (Shepherd, 1986). Length and thickness of hair were avoided by Laughery et al., and color does not lend itself easily to measurement. Clearly, any set of measurements used for mug-shot retrieval would have to include those relating to hair.

A second difficulty which geometric coding raises is the form in which the data should to be stored. If the database consists of measurements (raw or transformed for standardization), the input for a search would have to be in the form of such measurements. This implies that witnesses would have to generate sketches either by an artist or by facial composite system. Such images are unlikely to be of such precision as to allow comparisons on geometric indices to be made, as indicated by the fairly elaborate set of transformations Laughery and his colleagues had to make to prepare measurements from artists' sketches and Identikit composites. Furthermore, even if this problem is overcome, the "qualitative" attributes of faces such as hair color or style, or fatness of face, which are salient in the descriptions of witnesses would have to be coded in some other way.

An alternative is to use syntactic coding, in which everyday terms for describing faces such as "long nose," "wide-set eyes," and "dark hair" are used. Usually, as in Goldstein et al. (1972) and Lenorovitz and Laughery (1984), the descriptors are converted to a numerical value on a scale for coding, though this could be done via a program in response to qualifiers such as "very long," "average length," and so on. This has the advantage that the database lends itself to being searched on the basis of "natural language" descriptions, which can be converted to a similar set of ratings. The main disadvantage of the method for operational purposes, as mentioned above, is that coding large numbers of faces on a large number of attributes is time consuming, and for a single individual, prone to unreliability on quantitative attributes. Qualitative characteristics, such as color or texture, can be more reliably coded.

Once the database is established, the question of the search algorithm has to be decided. Again, Laughery et al. (1981) have drawn a useful distinction between "sequencing" and "matching" search algorithms. The sequencing algorithm is exemplified by Goldstein et al. (1972), as described above. A Euclidian-type distance measure, or goodness-of-fit, is established between the input and each of the cases in the database in turn, which can then be ranked according to this measure. The matching algorithm, as exemplified by Lenorovitz and Laughery (1974), uses each input parameter in turn to discard cases which fall outside the range of likelihood of acceptance so that only a small subset is eventually left. The latter procedure has the advantage of speed at the cost of risking the loss of the target through an erroneous parameter in the description. The former may take longer, but one error in the input can be outweighed by many more accurate ones, since the target is never rejected. It is an empirical matter as to what the respective costs of these disadvantages are.

A final question was the form in which images should be stored. Both Goldstein et al. (1972) and Laughery et al. (1971) stored their images as photographic prints. While this may be satisfactory for small sets of images, the labor involved in retrieving images from a large collection of many thousands stored in this way would be as great as that involved in creating albums of prints. Among possible storage methods are microfilm, microfiche, videotape, videodisc, and optical disc. Fiches and discs have the great advantage of rapid random access, and for this reason are more attractive.

FRAME—A PROTOTYPE MUG-SHOT RETRIEVAL SYSTEM

A prototype system for searching mug-shot files was developed at the University of Aberdeen. In designing FRAME (face retrieval and matching equipment), a series of decisions were made on the questions discussed above. First, the database should be in a form accessible through a witness' verbal description of a face. This meant syntactic coding of the images. Second, some means of geometric coding should be developed which could be translated into the form of the syntactic code. Third, a sequencing search algorithm should be adopted to avoid the risk of losing a target through a single error by a witness. And fourth, the images should be stored on a videodisc allowing computer-controlled random access to images.

Development of the Database

1. Ratings database. In its prototype form, FRAME comprised three images of each of 1,000 faces with a database of 50 parameters for each face. The original images were collected by photographing in color 1,000 men from Aberdeen in a special studio. Three poses were used, full face, semiprofile and full profile (see Shepherd, 1986, for more details). All faces were rated by a group of trained raters on 47 parameters from simultaneous projection of all three poses of every face. Three other parameters (age, height, and weight) were obtained from the subjects of the photographs. The full set of parameters is listed in Table 6.1. This comprised the Ratings database. The images were copied to an EMI videodisc, which was linked to a Cromemco System 3 computer and to a color monitor.

2. Physical measures. For operational purposes ratings of 47 parameters by a group of raters, or even by one person, is not practicable. To overcome this a semiautomatic procedure for measuring faces was adopted. Images were projected on to a bit-pad, and by locating specified points of the face with a stylus, coordinates of the points were defined. From these points, which are illustrated in Figure 6.1, a series of linear and area measurements was computed, and these measurements were correlated with ratings from the parameter list. From these correlations a set of measures corresponding to rated parameters was identified, and the ratings were regressed on to these measures to provide a parallel database of scalar values. Parameters for qualitative attributes remained unchanged.

Search Algorithm

A witness' description is used to search the database. In order to perform the search, the description is converted into a series of scale values for the relevant parameters for input. The search algorithm used for FRAME is a sequencing algorithm.

Since the database values are based on means, they are recorded as real numbers ranging from 1.0 to 5.0. Witness input is in the form of integers. The search algorithm therefore provides a variable tolerance range within which a matching score will be counted so that, for example, a database value which falls within plus or minus 0.5 of a scale point of the witness' input may be counted as a match. The program also allows for some of the parameters selected by the witness to be given a higher weighting in computing the matching score, which enables some account to be taken of the relative certainty of the witness about particular attributes.

Iteration of Searches

A witness may not retrieve the target high in the list of matches at the first attempt. Apart from errors in the initial description, the translation into scalar values for a search may be inaccurate if the witness adopts a frame of reference for quantitative judgments which is discrepant with that used in setting up the database. After the initial description the first 10 faces on the retrieval list are

Table 6.1.   Parameters on which faces were coded in the FRAME database. (Note—a P beside a parameter indicates that this parameter was subsequently coded by physical measurement—see text below for explanation.)

SHAPE OF FACE
1. short-long P
2. narrow-broad P
3. bony-fleshy

COMPLEXION
4. fair-dark
5. pale-florid
6. unlined-lined
7. clear-blemished

HAIR
8. short-long P
9. tidy-untidy
10. straight-curly
11. bald-full head P
12. no grey-white
13. black-brown-red-fair-blond

FOREHEAD
14. low-high P
15. narrow-broad P
16. straight-sloping

EYEBROWS
17. thin-thick P
18. straight-bent
19. meet in middle-wide set P
20. low-high P

EYES
21. small-large P
22. narrowed-open P
23. close set-wide spaced P
24. deep set-protruding
25. blue-grey-green-hazel-brown

EARS
26. small-large

NOSE
27. small-large P
28. short-long P
29. narrow-broad P
30. concave-hooked
31. small nostrils-large nostrils
32. narrow tip-broad tip P

MOUTH
33. small-large P
34. thin-thick upper lip P
35. thin-thick lower lip P

CHIN
36. small-large P
37. pointed-square P
38. receding-jutting

FACIAL HAIR (coded Yes/No)
39. none at all
40. moustache
41. sideburns
42. beard

PHYSICAL PECULIARITIES (Yes/No)
43. squint
44. bags under eyes
45. scars

ACCESSORIES (Yes/No)
46. glasses
47. earrings
48. Age
49. Weight
50. Height

displayed, and the witness, if he or she is unable to identify the target, can be asked to indicate in what way, if any, the retrieved faces differ from the target. As a result the witness may amend the values of the initial parameters, for example by observing that the hair of the retrieved faces is too long, or the nose too wide.

Alternatively, the witness may observe that one or more of the faces is similar to that of the target, in which case the database entry for the most similar face can be copied and entered as the search parameters for the target. The system thus provides the opportunity to iterate the witness' searches throughout the pro-

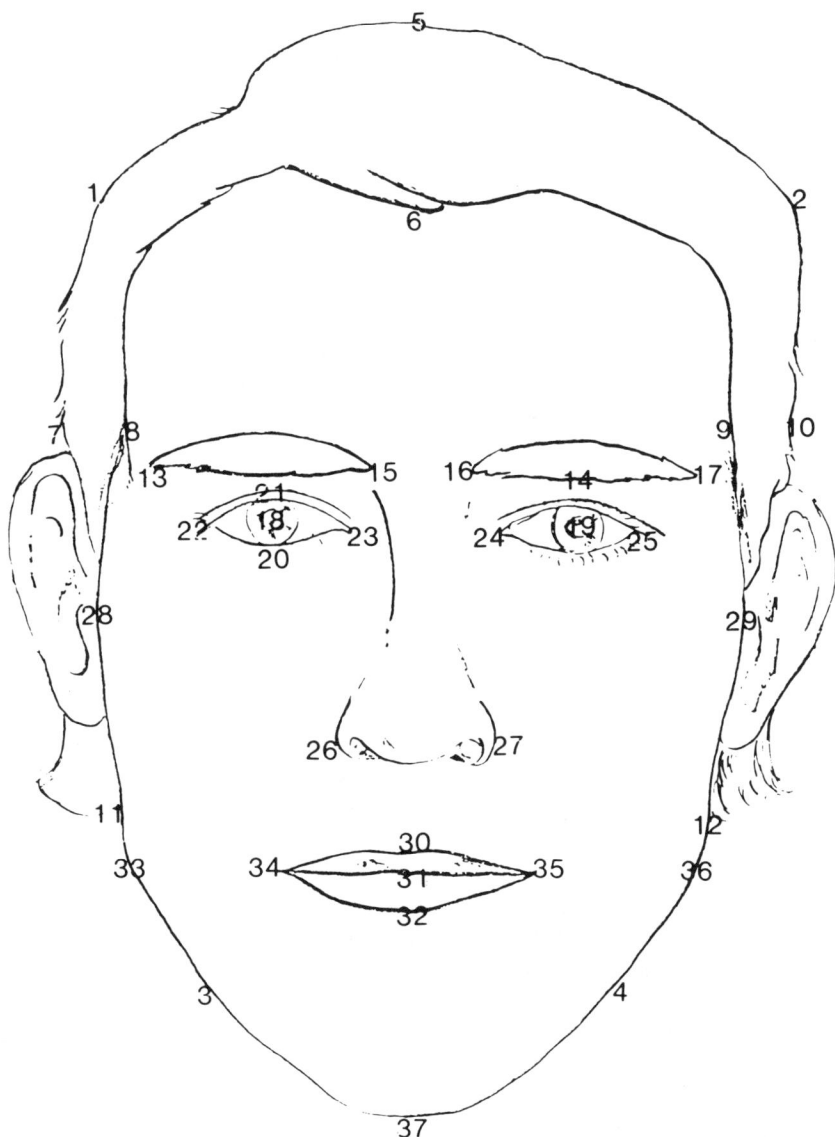

Figure 6.1. Points on the face used for measuring distances and areas in the Aberdeen FRAME system.

cedure, modifying descriptions on the basis of feedback from information entered earlier.

Initial Tests of the System

A series of laboratory experiments were run to test the effectiveness of the FRAME system. Some of these have been reported in Shepherd (1986) and Ellis et al. (1989) and will be only briefly summarized here. The principal experiment was a test of the efficacy of the FRAME system in comparison with the conventional mug-shot album procedure. Four distinctive (having a beard, moustache, glasses, or a bald head) and four nondistinctive targets were selected. Subjects were shown one of these targets and after a short interval were asked to describe him to an operator. In the FRAME condition, subjects went on to code their description using appropriate parameters, and this coded description was then used to search the database. The six faces with the highest matching scores were displayed, and if the target was not among these, the subject went through the iterative procedure described above. This was repeated for up to four searches, during which a maximum of 24 faces could be shown, and if the target had not appeared or was not recognized by the subject, the result was recorded as a miss. If the target appeared and was recognized, a hit was recorded. In the album method, the target was embedded in a collection of 1,000 faces arranged four to a

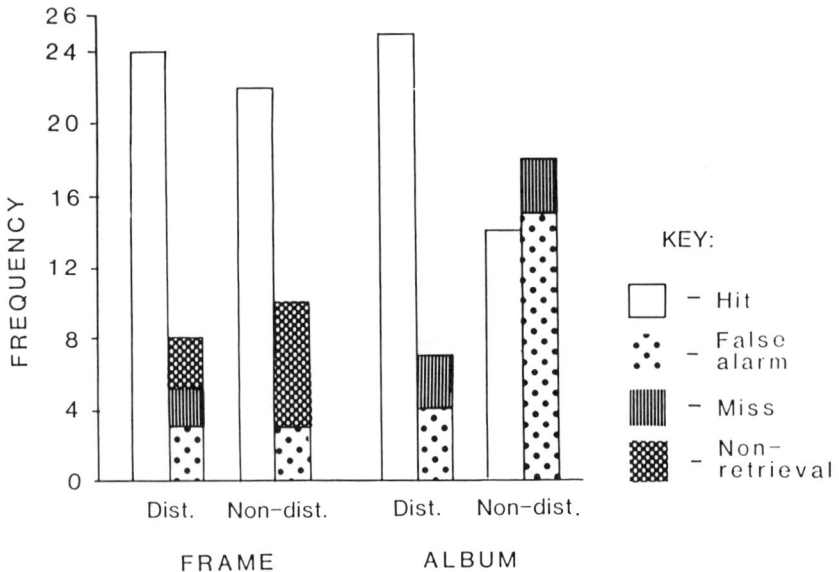

Figure 6.2. Hits and false alarms for distinctive and nondistinctive faces on FRAME and album searches.

page. The position of the target in the album was varied among four positions, 97th, 353rd, 639th, and 898th. Subjects in this condition could record a hit, a false alarm (false positive), or a miss (failing to select any face).

Results are illustrated in Figure 6.2. For the distinctive faces, hit rates for both FRAME and album were high at 75 and 78 percent, respectively. Results for the nondistinctive faces showed a clear superiority for FRAME with a hit rate of 69 percent compared with 44 percent for the album. A drop in the hit rate and a corresponding rise in false alarms in the album condition was strongly related to position, with the drop in performance being most marked when the target appeared in 898th position.

These results were very promising for the prototype. Hit rates of over 70 percent, where a hit is defined as the target occurring within the top 4% of retrievals, were typical for the laboratory trials of FRAME, and a similar success rate in the field would represent considerable improvement over current levels of performance by witnesses.

FACES SYSTEM

The initial laboratory tests of the FRAME prototype were followed by a field evaluation of the operational version FACES (Facial Analysis Comparison and Elimination System) which was installed in an English police force. The hardware comprises a microcomputer, framestore, microfiche system, vdu, digitizing pad, and color monitor, which are used for data capture, search, and system management.

The images are stored on microfiches mounted on a carousel, and are displayed on the monitor via a video-camera and framestore.

Data capture is executed using the digitizing pad for geometric coding, and by direct judgment for syntactic coding. The measurements for the geometric coding are obtained by using a stylus and pad to locate the 37 points identified in the FRAME project for facial measurement. Linear and area measures are computed and converted to real numbers on a 5-point scale ranging from 1.0 to 5.0, using regression equations derived from the FRAME database. The qualitative attributes are syntactically coded as discrete categories, and are estimated directly by the operator who enters an integer directly to the database. Data capture is thus semiautomatic for 21 attributes (indicated in Table 6.1), and requires operator judgment only for those attributes which can be reliably estimated.

The search procedure is similar to that used for FRAME except that it is menu driven. The initial free-description by the witness can be typed into a window at the top of the vdu screen and can be retained for reference. Appropriate parameters for the search can be selected by the operator on the basis of this description, and displayed singly for response by the witness, whose judgment may take the form of Yes or No for dichotomous categories such as presence of glasses, moustache, or beard; a verbal descriptor for discrete categories such as hair color or complexion which is entered as a numerical code; or as a rating on a continuous scale for scalar attributes. The last of these can be recorded by the witness controlling the movement of a cursor along a continuous scale to the point where he or she wishes to record his or her judgment. This enables a search parameter to be a real number, and does not restrict the entry for these parameters to integers.

The search algorithm and the facilities for revising initial values, varying the permitted range of tolerance, and weighting specific values are similar to those used in the FRAME prototype. However, FACES differs in the mode of displaying the "hit list." Single images are not displayed. Instead, a frame store retains the images as they are generated from the search, and these are subsequently displayed in sets of 12, randomized for order of retrieval within the 12.

At present the field trial is undergoing evaluation. Preliminary reports indicate a significant increase in the number of identifications from FACES searches compared to the usual rate for conventional mug-shot albums. What is more difficult to evaluate is whether the identifications are "true," since this depends on the subsequent arrest and disposition of the suspect. It is even more difficult to ascertain what proportion of cases is missed. A "miss" may occur for a number of reasons. The witness may fail to remember the suspect accurately, the suspect may have changed his or her appearance since data capture, and above all, the suspect may have no previous convictions, in which case he or she will not be in the database.

The use of a microfiche system for storing images has the advantage of ease of adding cases compared with the videodisc. However, optical discs are now in use by some police forces for storing images which are captured directly onto videotape and transferred to disc. This provides a large storage capacity in a small space, and linked to FACES could provide a powerful system.

FACE RECALL SYSTEMS

Mug-shot retrieval systems are invaluable when the suspect is on record. When there is no mug-shot of the suspect, or the investigators do not want to risk contamination of the witness' evidence which searching through mug shots might entail, the police often wish to release some visual image of the suspect's appearance, based on the witness' recall.

Artists' sketches and composite systems such as Identikit and PhotoFIT have long been adopted by police forces throughout the world. In recent years, computerized versions of these procedures have been developed, and some are in operational use (Shepherd & Ellis, 1990).

Before we address individual, computerized facial composite systems it is apposite to consider face recall from a psychological viewpoint. Compared with our ability to recognize faces, we are rather poor at recalling them (Ellis, 1984). There may be many reasons for this. For example, it may be that people can easily form a mental image of someone they have seen but simply find it difficult to externalize that image in a form that others can utilize. Phillips (1978) did find that subjects claimed to have good imagery shortly after exposure to a face, but the accuracy of such assertions is difficult to establish. Even if one can accept that good facial imagery is likely, however, it is clear that the problem of making public that image remains. Ellis (1986), in discussing such problems, appealed to the work of Kosslyn (1985), whose general model for mental images seemed appropriate for the specific, applied problem of trying to elicit from witnesses to a crime information about the appearance of a suspect.

Kosslyn identified four main aspects of the process:

1. Find (locating an image in memory).
2. Regenerate (maintaining the image).
3. Efficiency (ease of generating an image).
4. Detail (facility to "zoom in" on parts of the image).

Police witnesses may encounter difficulty with any of these stages. If they have few problems in finding, regenerating, or accessing details of an image, they still have difficulties in making public that knowledge.

In fact it is much more likely that witnesses have difficulty with one or more of the processes identified by Kosslyn (1985). In particular, there is some reason to believe that the main problem may lie with the inherent difficulty of zooming in on details of a face image to provide information on, say, size of eyes or width of mouth. It is difficult to provide any more than anecdotal evidence for this belief, but subjects in face recall experiments do sometimes spontaneously report just such problems, often adding that, as they attempt to read off details of specific facial features, they may lose the entire image.

When faces are imperfectly recalled, there is an interesting tendency for subjects to be able satisfactorily to provide information about face shape, hair, and age—which Ellis (1986) labelled as cardinal features. These are also the features that account for most of the variance in multidimensional scaling analyses of facial similarity judgments (Shepherd, Davies, & Ellis, 1981). One logical consequence of these findings is that, to be effective, facial composite systems must allow witnesses the opportunity to display their recollections of at least the three cardinal features. This means that facilities should exist for their easy manipulation. The most popular composite systems were Photofit and Identikit (Davies, 1981; Shepherd & Ellis, 1990). Neither of these allows such manipulation—or many configurational alterations that may be essential when tapping face memory (Sergent, 1984). To achieve an adequate range of both feature and configural manipulations computer imaging, display and control techniques seem essential.

COMPUTER COMPOSITE SYSTEMS

The first attempt to tackle the practical problems of face recall using a computer was made by Christie, Davies, Shepherd, and Ellis (1981). They compared the accuracy of conventional Photofit composites with those made on a computer system that used Photofit elements but also allowed witnesses to alter interfeature distances. However, despite the infinitely greater flexibility of the computer system, it did not produce more accurate composites than the Photofit system. This finding may have occurred as a result of limitations at the witness end which meant that no amount of technical sophistication can improve facial composites, or it may be attributable to the fact that the operators assisting the two groups of subjects had vastly different degrees of experience. We do know that operator experience is important in determining the accuracy of Photofit composites (Davies, Shepherd, Shepherd, Flin, & Ellis, 1986); and, therefore, this variable must not be overlooked.

The failure to show any advantage for a computerized facial composite system has not deterred others subsequently from trying to produce systems that will attract the attention of police forces throughout the world. We shall now give a detailed description of two of these new systems. These have been chosen to represent state-of-the-art techniques that are each representative of different approaches. E-FIT is an example of systems that harks back to Photofit by using multiple gray-scale representations of facial features. Comphotofit and Computer Facial Identification System (CIDS) also produce black-and-white "photographic" composites. Conjure and Kiwisoft Identikit produce full color composites. Mac-a-Mug Pro is a line-drawing or sketch composite system that echoes Identikit. It represents a group of systems that includes Compusketch and Sigma.

EFIT(Electronic Facial Identification Technique)

Unlike most other electronic composite systems, the image produced by EFIT approximates photographic quality rather than a line-drawing image. Empirically there are no data on which to judge specifically which kind of system is operationally most effective. Previous research using photographs of a target in its original form and line tracings from the photograph of varying detail has found that the photographic quality image results in superior recognition (Davies, Ellis, & Shepherd, 1978). However, in this case the photograph was a veridical image of the target, and not a witness' construction. Other studies have compared sketch artists with Identikit line drawings (Laughery & Fowler, 1980) and PhotoFIT with a computerized version of the same system (Christie et al., 1981), but the evidence from these does not bear directly on the issue.

On the evidence of Davies et al. (1978), it might be suggested that, where a comparable degree of verisimilitude is obtained from both a photographic and a line-drawing system, the photographic image should lead to superior recognition and should have more aesthetic appeal.

EFIT was developed jointly by the Home Office, Io Research Ltd., and the Psychology Department of Aberdeen University. Its hardware comprises a personal computer with 70Mb hard disk and text screen, a color monitor, and a Pluto graphics unit, with an optional graphics tablet and stylus. The software consists of a feature library, an index for accessing the features, and a program for assembling and modifying images on screen. An optional additional paint package is also available.

The features in the caucasian feature library include 141 general face shape (including chin), 440 hair, 258 eyes, 123 eyebrows, 204 nose, 200 mouth, 78 ears, and 70 of various forms of facial hair. In addition, there are a variety of accessories including scars and tattoos, hats, glasses, and so on. A noncaucasian feature library is under development. Features are stored separately, having been cut individually from photographs and edited for tonal matching.

They are indexed by sets of descriptors which have been empirically derived in research at Aberdeen and elsewhere from everyday descriptions of faces, and are similar in content to those used for FRAME. Once the features have been described by the witness, the completed composite is displayed, a "default" feature being used where a witness is unable to provide a description.

The completed composite can be modified by replacing any or all of the features, or by changing the shape, position, or tone of any features. There is also a facility for coloring the image. Figure 6.3 illustrates some of the changes which can be achieved with the most basic version of EFIT.

In its basic form, EFIT is driven by a series of menus and no artistic ability is required. However, the availability of a full paint system permits extensive artwork on any image, and the creation of "unique" features such as bizarre hair styles, tattoos, etc.

The availablity of an image-grabbing facility enables an operator to create on a video display any scene or object for which he or she has an image. For example, the scene of a crime may be photographed, captured by the imager, and specific aspects (for example, model of car) put in position to recreate the scene for the witness.

Mac-a-Mug Pro ™

Mac-a-Mug Pro is a line-drawn male, caucasian facial composite application for Macintosh computers. It comprises 185 hairlines, 118 eyes and eyebrows, 14 ears, 66 noses, 81 mouths, and 46 chins. Features may be selected by number from a manual or, can be addressed using a mouse on analogue scales ranging, roughly, from fine to coarse. When a witness has made his or her first selections,

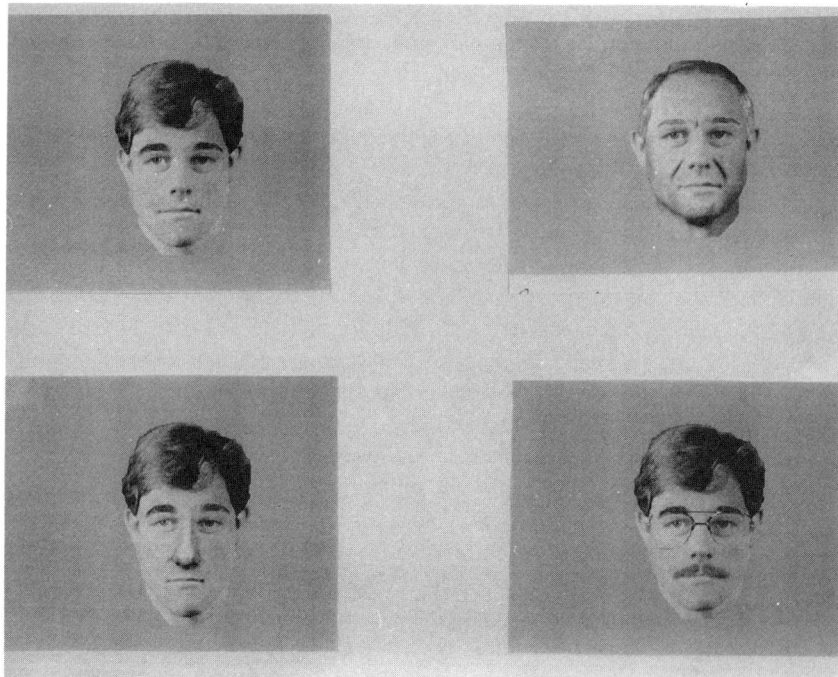

Figure 6.3. Examples of manipulations to a facial composite using EFIT. Top left: original composite. Top right: "Aged" face—inner features unchanged from original; new face shape, hair, and added brow and nose lines; original eyebrows thinned and lightened. Bottom left: Original face shape lengthened, original lips made thinner, original mouth narrowed, new nose. Bottom right: original face with moustache and glasses added.

they may be quickly replaced by choosing other features or the configuration may be altered by moving selected features up, down, left or right in 1-, 5-, or 10-pixel steps.

Mac-a-Mug Pro also contains accessories such as hats, spectacles, moustaches, and beards, and faces may be darkened to produce noncaucasian faces. Tools exist to make alterations freehand or by spraying. In addition, composite faces, for example, may be made narrow or broader by importing the file to other graphic applications such as Superpaint. Figure 6.4 gives some examples of how alterations may be made to a Mac-a-Mug Pro composite.

It almost goes without saying that Mac-a-Mug Pro has not been objectively evaluated by its publishers (Shaherazam)—or yet by anyone else. Figure 6.5 indicates the potential of Mac-a-Mug Pro. The Woody Allen composite made from a photograph is acceptable but much remains to be done before one may conclude that Mac-a-Mug Pro has good potential for forensic use.

Target face

New nose

New eyes +eyelines.

New hair

Dark features,=negroid

Female face

Figure 6.4. Examples of manipulations to a facial composite using Mac-a-Mug.

Figure 6.5. A construction of Woody Allen using Mac-a-Mug.

Indeed, at present there are no empirical grounds for preferring a "pho-tographic" approach or a "sketch" approach for enabling witnesses to recall suspects' faces. Each has prima facie advantages: the photo methods provide composites that police forces seem to prefer because of their lifelike qualities; the sketch methods may be more effective precisely because they are less realistic and can, therefore, more easily convey a type likeness (which may be all that most witnesses are capable of recalling). The time is ripe for a formal scientific comparison of, say, E-Fit with Mac-a-Mug Pro, in order to asses the general advisability of one approach against the other. At the moment, however, none of those with vested interests seems to be very keen to do so.

REFERENCES

Batten, G.W., & Rhodes, B.T. (1978, May 17–19). *UHMFS. The University of Houston Mugshot Filing System.* Paper delivered to the Carnahan Conference on Crime Countermeasures, Lexington, KY.

Christie, D., Davies, G.M., Shepherd, J.W., & Ellis, H.D. (1981). Evaluating a new computer-based system of face recall. *Law and human behavior, 5,* 209–218.

Davies, G.M. (1981). Face recall systems. In G.M. Davies, H.D. Ellis, & J.W. Shepherd (Eds.), *Perceiving and remembering faces.* London: Academic Press.

Davies, G.M., Ellis, H.D., & Shepherd, J.W. (1978) Face recognition accuracy as a function of mode of representation. *Journal of Applied Psychology, 63,* 180–187.

Davies, G.M., Shepherd, J.W., & Ellis, H.D. (1979). Effect of interpolated mugshot exposure on accuracy of eyewitness identification. *Journal of Applied Psychology, 64,* 232–237.

Davies, G.M., Shepherd, J.W., Shepherd, J., Flin, R., & Ellis, H.D. (1986). Training skills in police photofit operators. *Policing, 2,* 35–46.

Ellis, H.D. (1984). Practical aspects of face memory. In. G.Wells & E. Loftus (Eds.), *Eyewitness testimony.* Cambridge, England: Cambridge University Press.

Ellis, H.D. (1986). Face recall: A psychological perspective. *Human Learning, 5,* 189–196.

Ellis, H.D. , Shepherd J.W., Shepherd, J., Flin, R.H., & Davies G.M. (1989). Identification from a computer-driven retrieval system compared with a traditional mug-shot album search: A new tool for police investigation. *Ergonomics, 32,* 167–177.

Geiselman, R.E. , Fisher, R.P., MacKinnon, D.P., & Holland, H.L. (1986). Enhancement of eyewitness memory with the cognitive interview. *American Journal of Psychology, 99,* 385–401.

Goldstein, A.J., Harmon, L.D., & Lesk, A.B. (1972). Man-machine interaction in human-face identification. *The Bell System Technical Journal, 51,* 399–427.

Kosslyn S. M. (1985). *Ghosts in the mind: Creating and using images in the brain.* New York: Norton.

Laughery, K.R., Alexander, J.F., & Lane, A.B. (1971). Recognition of human faces: Effects of target exposure time, target position, pose position, and type of photograph. *Journal of Applied Psychology. 51,* 477–483.

Laughery, K.R., Duvall, G.C., & Fowler, R.H. (1977). *An analysis of procedures for generating facial images* (Mug File Project rep. number UHMUG-2). University of Houston, Texas.

Laughery, K.R., & Fowler, R. (1980). Sketch artist and Identikit procedures for recalling faces. *Journal of Applied Psychology, 65,* 307–316.

Laughery, K.R., Rhodes, B., & Batten, G. (1981). Computer-guided recognition and retrieval of facial images. In. G.M. Davies, H.D. Ellis, & J.W. Shepherd (Eds.), *Perceiving and remembering faces.* London: Academic Press.

Lenorovitz, D.R., & Laughery, K.R. (1984). A witness-computer interactive system for searching mug files. In G.R. Wells & E.F. Loftus (Eds.), *Eyewitness testimony. Psychological perspectives.* Cambridge, England: Cambridge University Press.

Phillips, R.J. (1978). Recognition, recall and imagery of faces. In M. Gruneberg, P. Morris, & R. Sykes (Eds.), *Practical aspects of memory.* London: Academic Press.

Sergent, J. (1984). An investigation into component and configurational processes underlying face perception. *British Journal of Psychology, 75,* 221–242.

Shepherd, J.W. (1986). An interactive computer system for retrieving faces. In H.D. Ellis, M.A. Jeeves, F. Newcombe, & A. Young (Eds.), *Aspects of face processing.* Dordrecht, Netherlands: Nijhoff.

Shepherd, J.W., Davies, G.M., & Ellis, H.D. (1981). Studies of cue saliency. In G.M. Davies, H.D. Ellis, & J.W. Shepherd (Eds.), *Perceiving and remembering faces.* London: Academic Press.

Shepherd, J., & Ellis, H.D. (1990). Systeme zum Abruf von Gesichtsinformationen [Face recall systems]. In G. Köhnken & S.L. Sporer (Eds.), *Identifizierung von Tatverdächtigen durch Augenzeugen.* Stuttgart, Germany: Hogrefe.

7

The Use of 3-D Computer Graphics for the Simulation and Prediction of Facial Surgery*

Alfred D. Linney
Department of Medical Physics
University College London

INTRODUCTION

With recent advances in anaesthesia and surgical technique, maxillo- and cra-niofacial surgery has now become so sophisticated that surgeons can alter the shape of most of the components of the facial skeleton and forehead (Burke, Banks, Beard, Tee, & Hughs, 1983). With this capability, acceptable facial harmony has increasingly become the goal of surgery rather than simply correcting jaw disproportion and malocclusion. In spite of this, surgical planning for the correction of facial deformity has largely been based on the analysis of lateral skull radiographs which allow only profile changes to be confidently planned (Epker & Fish, 1986). The patient, who rarely sees his or her profile, is naturally more concerned with full face appearance.

Since the planning of reconstructive maxillofacial and craniofacial surgery is in reality a problem of the rearrangement of the anatomy in three dimensions, the postsurgical appearance of the full face can only be predicted using three-dimensional modelling. In the planning of complex facial operations, plaster and wax models are often constructed so that the three dimensional relationships of

---

* The work described here has been very much an interdisciplinary team effort. Professor James Moss has pioneered the clinical use of surgical simulation by the use of computer graphics. Simon Arridge and Sue Grindrod have developed the necessary programming techniques to turn an idea into a reality. In the course of development, funds have been gratefully received from the University College Hospital Special Trustees and the DHSS Information Technology Division. Lastly, I wish to thank Aurea Fenty for patiently typing and correcting the manuscript.

deformed facial and cranial bones can be appreciated, and proposed surgical corrections can be rehearsed and evaluated. The techniques for planning surgery, particularly in the cases of facial asymmetry, are now regarded as lagging behind the actual surgical procedures themselves (Hall, 1985).

The quest to establish a quantitative basis for planning alterations to facial appearance and skeletal function by orthodontic and surgical intervention has a long history. From the earliest developments of these techniques it was realized that accurate measurements of the face, the arrangement of the teeth, and the underlying facial bones were of vital importance. A number of mechanical devices for the measurement of points on the facial surface have been developed (Hall, 1985), and the detailed analysis of plaster casts of dentition have also figured strongly in orthodontic practice. Two-dimensional projections in the form of plane X-rays of the head have until recently been the most important source of information for the analysis of the underlying bone structure for individual patients. This has led to the establishment over a period of years of a highly consistent and widely accepted system for taking the X-rays, and to a number of ingenious attempts to make maximum use of this limited and sometimes ambiguous data.

For example, as part of a much larger endeavor to develop a universal method for analyzing facial shape, Rabey (1977) produced a system which allowed X-rays of the skull to be accurately superimposed on photographs of the face and upon analytic lithographs from both the lateral and full face view. In this way, the relationship between the soft tissue and the underlying bone could be determined. Using this technique, information could be derived which allowed an indication of the effects of surgery on the soft tissues and hence possibly facial appearance to be obtained. A less elaborate but apparently quite effective idea has recently appeared in the literature using the superimposition of radiographic tracings and 35-mm photographic slides of the face to produce a sketch of the postsurgical facial appearance (Kinnerbrew, Hoffman, & Carlton, 1983) for both profile and frontal views. This technique, however, relies heavily on the personal skills and artistry of the surgeon using it. Cutting, Grayon, Bookstein, Fellingham, and McCarthy (1986) have also described a method of obtaining three-dimensional landmark locations from a combination of posteroanterior and lateral skull radiographs. A simple visualization of the gross pattern of the bones is then obtained by displaying the lines joining these points on a video screen. The method then uses this three dimensional constellation of points to optimize surgical planning, by determining which surgical procedures best bring them into what is considered to be a normal spatial relationship.

Very soon after computers became more generally available, they were employed in an attempt to reduce the manual drawing and computational burden and to provide extra information from the analysis of lateral skull radiographs. Computers were also used (Burke et al., 1983) for the analysis of stereo pairs of facial photographs from which horizontal and vertical profiles of the facial surface were derived. These were then used to evaluate changes brought about by surgery on the lower jaw.

A number of computer systems which allow and extend the traditional 2-D analysis of X-rays of the skull are now available (Bhatia & Sowray, 1984). These systems facilitate the automatic measurement of the relative positions of landmarks identified and marked by an operator. The traditional evaluation of angles between cranial reference planes is now performed by computer, and using this modest amount of information, the necessary adjustments of landmarks required to produce accepted norms for occlusion and the configuration of facial bones are computed. Although of undoubted value, the application of computers to the traditional analysis and surgical planning does not overcome its main limitation, which is the lack of three-dimensional information. There is also no adequate representation of the aesthetic adjustments to the face following the normalization of functional relationships. Any prediction of soft tissue response is confined to the mid-line profile of the face.

Although the importance of the third dimension in surgical planning had been appreciated for a long time, the main difficulty facing surgeons was that of obtaining sufficient three-dimensional measurements on the face and skull. Over the last decade, this situation has changed dramatically, and large numbers of measurements with an accuracy better than 1 mm are now rapidly produced over both the facial surface and the bones of the skull.

The introduction of the medical imaging technique known as *computerized axial X-ray tomography* (universally referred to as CT scanning) has been the most important development in this respect (Hounsfield, Ambrose, Perry, & Bridges, 1973). In fact, this system has received unparalleled attention in the history of medical imaging since its introduction, and has brought a Nobel Prize to Hounsfield, its inventor. The data on the human anatomy produced by this device come, not immediately as a film or picture, but in the form of an array of numbers representing a slice of the body. Each number is a measure, expressed in Hounsfield units, of the ability of the tissue found within a small volume element within the body slice to absorb X-rays. These volume elements typically measure less than 1 mm square and have a length equal to the slice thickness, which is also of the order of a millimeter or so. Where bone is present the corresponding array element contains a high number, and where an air cavity or fat is found, a low number appears. Other tissues are represented by intermediate values. Computer display techniques are used to produce images form these arrays, and a typical CT scan through the head is shown in Figure 7.1. Here each pixel (picture element/cell) is assigned a brightness proportional to the Hounsfield number. In this scan taken through the lower jaw, bone appears white, air black, and the soft tissues are displayed at intermediate shades of grey. Thus, unlike the traditional X-rays, in which the whole anatomy is projected onto a single plane, the CT scan images give a true three-dimensional picture of the anatomy contained within an array of tiny volume elements over a particular slice of the body. Moreover, since

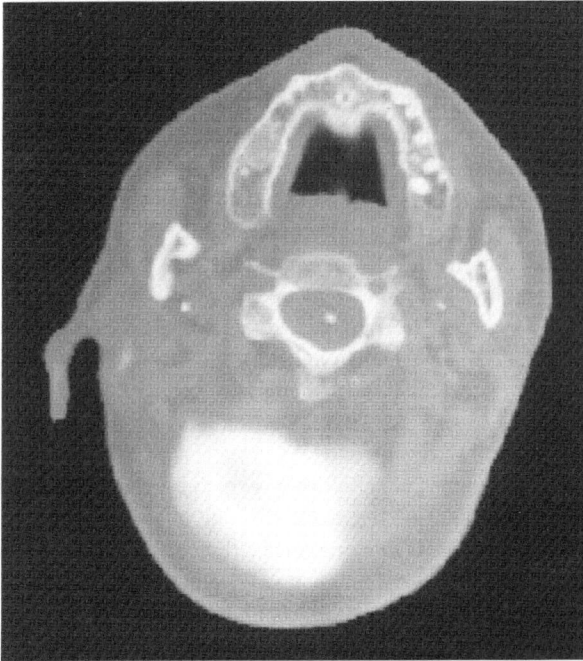

Figure 7.1. Axial CT scan through a patient's head at the level of the upper jawbone.

the data is produced in numerical form, image-processing techniques may be directly applied to enhance and separate out various anatomical components and to identify the boundaries between them. Some of these boundaries, for example, that between skin and air, can be found automatically by the computer (Figure 7.2), while some of the less obvious boundaries must be interactively marked by the operator. For most studies, and especially where an insight into the three-dimensional anatomy is required, a number of adjacent scans are taken. Because of their axial orientation, however, a great deal of skill, usually that of a radiologist, is required to reliably trace and report on clinically significant features extending across several scans. Even then, the 3-D image exists only in the mind of the radiologist, and no permanent reproducible record is produced.

The CT slices alone present considerable difficulties in conceptualization when applied to the facial bones, and more particularly in abnormal facial structure, where the anatomical cues usually used to guide interpretation may be absent. In fact, no planar format is completely satisfactory in demonstrating the complex morphology of the facial bones and their interrelationship. The simplest ways of improving 3-D conceptualization is by allowing a set of sequential CT slices to be viewed together from any angle in their correct geometrical relation-

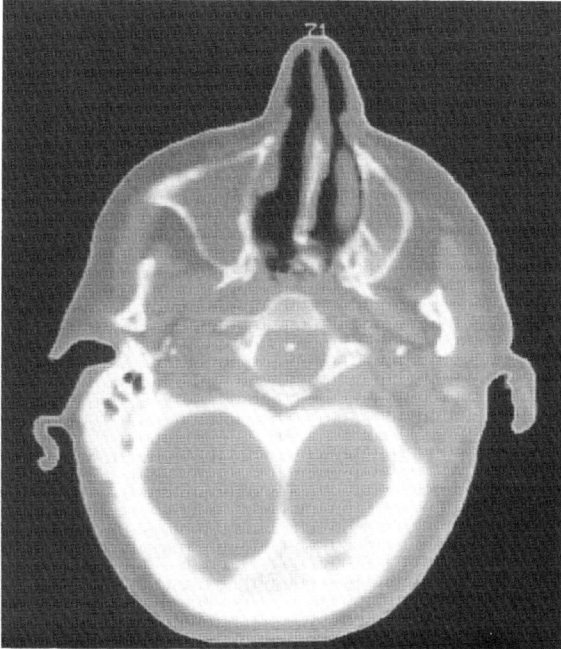

Figure 7.2. Automatic generation of skin surface profile.

ship or by showing a series of slices one after the other in rapid sequence to produce a kind of cinematographic effect. These methods of presentation, however, do not give a clear impression of the surfaces by which the nonradiologist usually perceives anatomical structure. A more explicit and useful method of displaying the three-dimensional anatomical surfaces came with the advancing techniques of computer graphics (Newman & Sproull, 1981), which had steadily become more sophisticated over the period in which CT scanning was coming into widespread use. In this field methods had been developed for producing realistic images of solid objects with sufficient visual cues to allow the viewer to perceive their three-dimensional character. The most important cue was provided by exploiting realistic surface shading to enhance the perception of surface orientation (Ramachandran, 1988). Gabor Herman and his colleagues showed that it was possible to reconstruct a surface from a set of CT scans and display this with the necessary 3-D cues, and was able to report on the visualization of internal organs using CT scan data and computer graphics techniques (Herman & Liu, 1979).

Since the different classes of tissue (i.e., bone, fat, muscle) have quite different CT numbers, it was also possible, by imposing a range limit on the CT data used, to display bone surfaces alone with the soft tissue stripped away. The possibilities of using this facility for the analysis of facial surgery soon became apparent, and in 1983, Vannier, Marsh, and Warren (1983) reported on the analysis and evaluation of a variety of surgical corrections for craniofacial disorders and presented several case histories in detail. The graphics described had been used effectively in the management of over 200 patients. Significantly, their 3-D graphics system had been used in a limited sense for planning the surgery as well as in evaluation. In some cases a *computer-aided design* (CAD) system, which the McDonnell Douglas Corporation (St. Louis, MO) had developed for the design of aircraft, was used. This system allowed life size models of the individual patient's skull to be produced by use of a numerically controlled machine tool. Vannier and his colleagues reported that the use of CAD has proved particularly effective in solving the difficult problem of determining the translocation required to bring the left orbit of the eye of a patient suffering from asymmetric hypertelorism to normalcy.

It soon became clear, however, that there were special problems associated with the use of computer graphics for the planning of surgical procedures which could not be solved by standard CAD techniques. The main differences arose from the amount and nature of the data to be handled and the speed with which results needed to be produced for clinical use (Flynn, Matteson, Dickie, Keyes, & Bookstein, 1983). The idea, however, that computer-aided 3-D analysis and simulation of surgery could permit the surgeon to rehearse a wide variety of surgical plans in detail preoperatively, and then to select the optimum surgical plan for each patient according to quantitative criteria, had clearly been established. It was the attractiveness of this which led to groups in both the U.S. (Brewster, Trivedi, Tuy, & Udupa, 1984; Barrett, 1984) and the U.K. (Arridge, Moss, Linney, & James, 1985; Moss, Grindrod, Linney, Arridge, & James, 1988) to begin to develop systems based on the use of 3-D graphics which would meet the planning needs of maxillofacial surgeons. The principles on which these are based necessarily differ from those of CAD systems, and the medical application of 3-D computer graphics has slowly become a subject of study in its own right.

At University College London a collaborative program was established in 1978 between the Medical Physics and Orthodontic Departments for the development of a system based on these principles for the simulation, prediction, planning, and analysis of facial surgery. The first stage was the construction of a laser scanner for recording the geometry of the facial surface. Later, the implemention of the 3-D graphics program was funded by the Information Technology Division of the Department of Health and Social Security. The techniques used, and the facilities developed, will be presented in some detail. The computer used was a NORSK 540D, driving a GEMS display system, the choice being made on the basis of an evaluation of suitable combinations of a general purpose computer and the display systems available at the beginning of main development phase (1984). Since the graphics databases are very large (of the order of 16Mbyte) and the number of computations required is enormous, both a large memory capacity and speed were the main requirements. A suitably large mass storage system and magnetic tape backup system was also needed to store patient CT scan data and laser data, and to archive selected images.

METHODS FOR USING 3-D COMPUTER GRAPHICS FOR SURGICAL PLANNING

The procedure for the simulation of surgery using computer graphics is shown schematically in Figure 7.3. Four separate stages may be identified.

1. Data Acquisition
2. Data Representation—formation of graphics database.
3. Display—presentation
4. Interactive modification of the images—surgical simulation

These steps will now be described with particular regard to their limitations and difficulties.

Data Acquisition and Preparation

Data acquisition for 3-D simulation of facial surgery requires the use of a CT scanner which produces data from which the bone and soft tissue structures can easily be extracted.

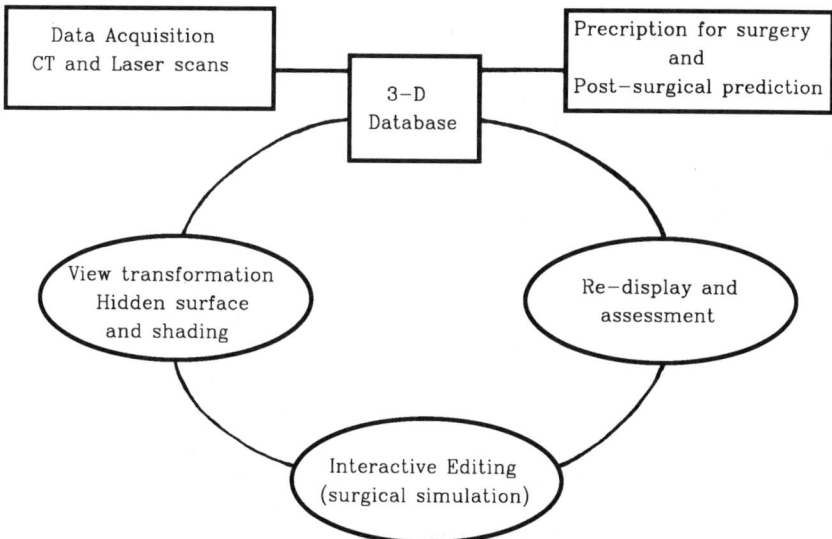

Figure 7.3. Overall scheme for the simulation of surgery.

Because of the X-ray dose, the CT scans must be carefully planned to give sufficient detail where it is needed, but to avoid overscanning regions of less importance which are only required for the sake of overall anatomical orientation. A suitable protocol for scanning the head has been established. With scan beam widths of 1.5 mm, scan spacing over the volume on which interest is focused may be either 1.5 mm or 3 mm, and 6 mm or more over the less critical volumes. This inevitably means that no data is collected on tissue volumes between the scans, a fact which presents some problems in the representation and display stages. Another quite serious problem for the scans containing the teeth and jaws is the common presence of amalgam or gold fillings and metal implants or prostheses. These cause highly undesirable streak artifacts in the CT scan, and there is at present no real alternative but to edit these features out by interactive manual intervention. The operator requires a reasonable knowledge of local anatomy to achieve this. The most likely source of error in the acquisition of multiple CT scans would be patient movement between scans, and, to prevent this, the patient's head is restrained during scanning. Geometrical errors are quite small, the geometry of most scanners being apparently accurate to within at least one millimeter.

Recently, Gholkar, Gillespie, Hart, Mott, and Isherwood (1988) have demonstrated that it is possible to perform a rapid sequence of scans delivering a much lower dose of X-rays, and which produce good data for bone imaging in three dimensions. Both the problem of patient movement and the radiation hazard involved in multiple CT scanning may therefore be greatly reduced.

Although CT data alone would be all that is required to build a graphics database for initial planning and simulation, another method of accurately monitoring changes in the facial surface postsurgically is essential for follow up. CT is not regarded as suitable for this purpose because of the cumulative X-ray dose which would be incurred. A second method is therefore required which does not involve the X-ray hazard but gives sufficient information for a quantitative understanding of the postsurgical changes in facial form. Optical methods may be used, and a laser scanner has been designed and built at University College London (Moss, Linney, Grindrod, Arridge, & Clifton, 1987) for this purpose.

The laser scanner uses trigonometric principles to determine the shape of the face by measuring the apparent distortion of a line projected onto the face when viewed at an oblique angle by a video camera. The method is illustrated in Figure 7.4. The chair in which the patient sits is rotated under computer control so that the whole facial surface may be measured. The distortion of the laser line as it passes over the face is recorded using a CCD video camera coupled to the computer system via a purpose built signal preprocessor and interface. The coordinates of approximately 20,000 points on the facial surface are automatically measured in 30 seconds, and the data are transferred directly to computer memory. The procedure imposes no hazard to the patient, but only surface data are produced.

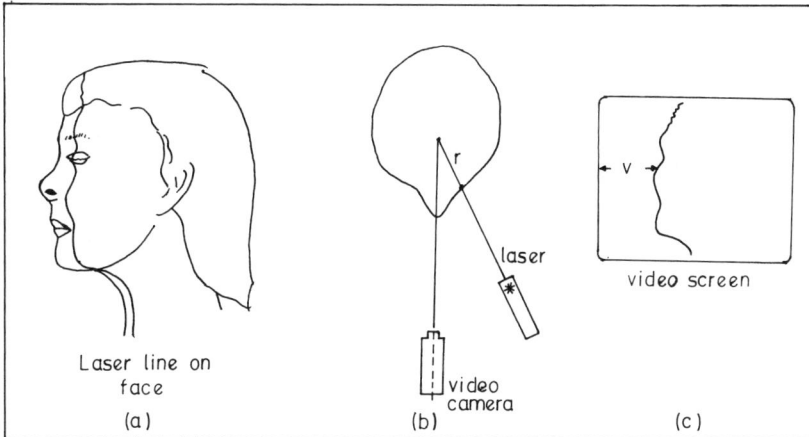

Figure 7.4 The UCL laser scanning system.

Data Representation

The acquired data have now to be converted into a graphics database for immediate use by a graphics display program. The types of database presently constructed for medical graphics have been dictated to a large degree by considerations of the requirements of the display algorithms themselves, the fastest algorithms generally being the most favored. In cases where special purpose computing hardware is used, its architecture and processing modes will determine database structure (Goldwasser, Reynolds, Talton, & Walsh, 1988). For the simulation of surgery two types of database have been found to be essential, one representing surfaces only and the other representing volume information. The surface type database which is used to represent the face can be derived directly from the profiles measured by a laser surface scanner or by the use of contour following from CT scan data. The volumetric database is used to represent the complex anatomy of the skull and is derived from CT scans only.

Clearly, for an effective simulation of surgery, the database used must contain information on the distribution of bone and soft tissue over the volume of the patient's head, and the most effective volumetric representation of complex anatomical objects has been obtained by dividing the space containing the object into cubic volume elements called *voxels* (Herman & Liu, 1979). For the head and skull these will usually be 1 mm in size. The ease with which the voxel array may be spatially subdivided into smaller groups of voxels also makes it highly suitable for the solid sectioning required for the simulation of surgical procedures.

It would be convenient to have a single database structure, but unfortunately, fast, simple graphics algorithms using the voxel database do not usually produce very realistic images of smooth curved surfaces, but rather images still showing signs of the discrete block structure (Herman & Liu, 1979). For presenting the human face, such a textured surface is unacceptable, and it is necessary to use a surface-based method, where only the surface itself is represented, usually as a series of connected tiles called *facets* which are most often triangular in shape (Moss et al., 1987) (Figure 7.5). More sophisticated shading algorithms can be used on faceted databases, giving a better sense of realism to the displayed surface. The facets are derived automatically from the CT or laser contour data using an "autofaceting" program which determines which data points are to be joined up to form each facet. Approximately 30,000 facets are formed from the facial coordinates produced by the University College laser scanner.

The generation of a voxel database from the CT data involves more computing procedures and can require some tens of minutes of computer time, depending on the structure of the database used and the completeness of the information stored.

Let us now look in more detail at the manner in which the voxel database is created from a series of CT scans. The scan data already come in the convenient

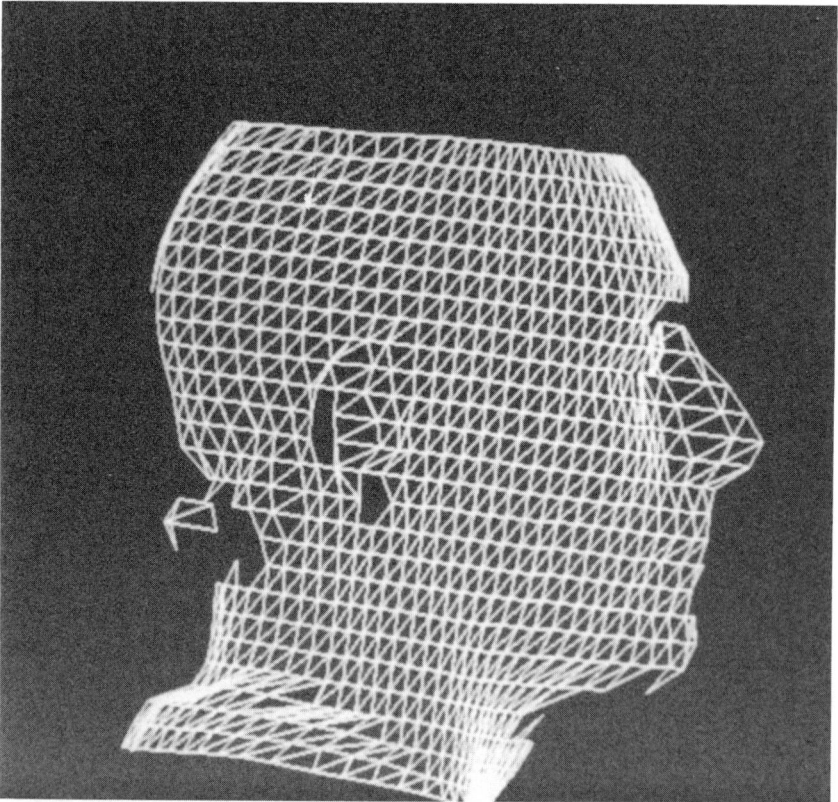

Figure 7.5. Representation of the face as a facetted surface.

form of arrays, each array element representing the X-ray absorption properties of approximately a 1–2 mm cube of human anatomy. It would thus seem very convenient to take a CT scan at millimeter intervals over the anatomical volume of interest, but this many scans would deliver an unacceptable radiation dose (Norwood, Cunningham, Bowseley, & Taylor, 1988) to the patient, and under the protocol adopted for data acquisition, scans are taken at larger intervals, with scan spacing varying between 1.5 and 6 mm.

Thus because of the need to keep the X-ray dose to patients as low as possible, the CT scans are generally separated by a larger distance than the size of the pixels within the scans themselves. To fill the gaps between scans and to produce the cubic voxels favored for the graphics database, data is assigned by interpolation to the "missing" 1-mm slices from the measured values on either side of them. The attempt, however, to reconstruct missing data from discrete samples in this way has many pitfalls, and some artefacts appearing in the eventual 3-D image are due to this procedure.

The most important of these artefacts arises where thin bones lie nearly parallel to the CT scan plane. In this case, the voxels have low values compared with that expected for bone, since they are only partially filled with bone, and may not be included when bone is selected from the data by separating voxels with a given range of CT values (windowing). The interpolated voxels will be likewise affected, since, where the adjacent scans are either side of a transition between bone and soft tissue, all of the interpolated values will give partially filled voxels. In these areas bone will disappear in the 3-D images, leading to the appearance of holes, or "pseudoforamina," in the displayed bony anatomy (Tessier & Hemmy, 1986). This effect is particularly pronounced for the floor of the orbits of the eyes and at the base of the skull, as shown in Figure 7.6. The effect is very difficult to remove entirely, since, even if information is used on the position of thin bones in the normal anatomy to enhance their detection, these will not necessarily be present in the abnormal or damaged skull. In such cases, procedures for removing pseudoforamina carry the risk of creating bone where there is none. In cases however, where fractures are to be analyzed, an awareness of the possible presence of pseudoforamina is vital.

One other database facility is included as essential for any system for the simulation of surgery. This is a program for conversion between the voxel and surface type database, so that the surgery simulated using the former can be translated into surface changes represented by the latter. This is particulary necessary for the presentation of images of the predicted postsurgical face.

Display

Several procedures are necessary for the generation of 3-D images whichever database is used. These are transforming the data for a selected viewpoint,

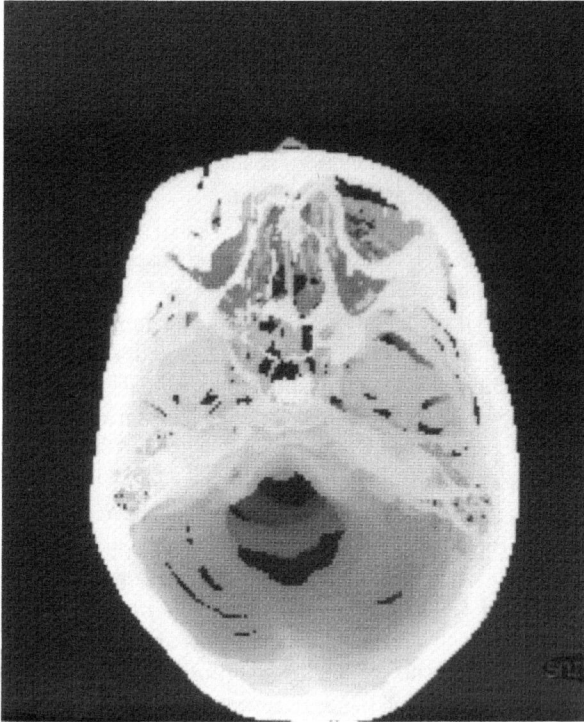

Figure 7.6. 3-D depth shaded image of the skull from superior view showing the presence of pseudoforamina.

determining which parts of the object in terms of facets and voxel faces will be visible from the chosen viewpoint, and shading the visible surface in such a way as to give strong cues as to its three dimensional character (Newman & Sproull, 1981).

For both the voxel and surface facet representations the problem of hidden surfaces is quite often solved by using a technique known as the *painter's algorithm*. Using this method, first projections of the surfaces furthest away from the observer are painted onto the screen; the next nearest ones are then painted over them, until finally the image is finished by painting those closest to the observer. The hidden surfaces are thus naturally obscured in the final image, in the same manner as a painter would precede by laying down the sky and background on the canvas before painting buildings and lastly people and objects in the foreground. This is the technique used for the faceted images in this chapter.

Although this algorithm works well, it is not among the fastest-hidden surface techniques, and since speed is an important factor in planning surgery, other methods have been explored. One such method is known generally as the *front-*

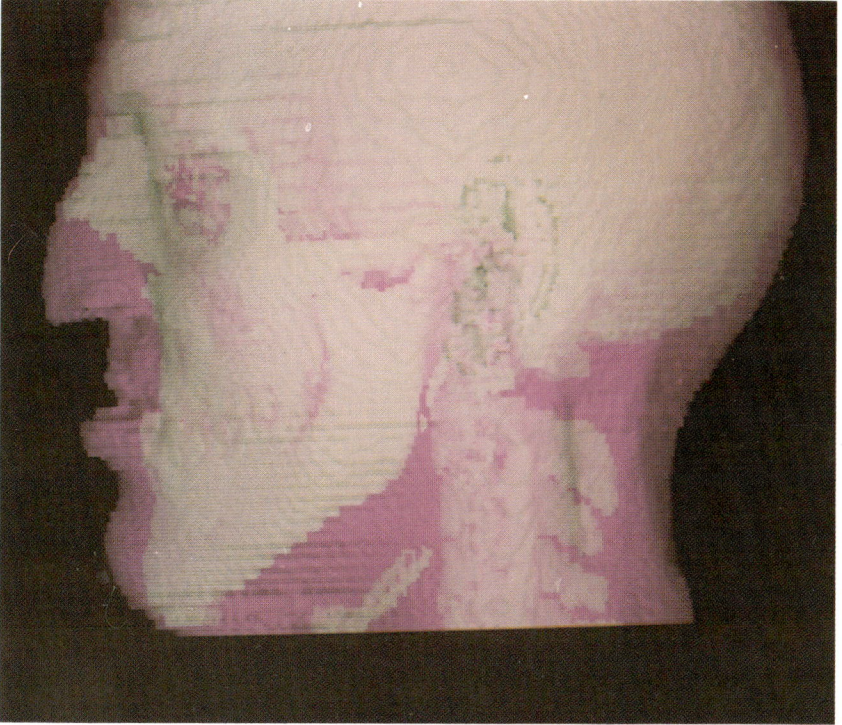

Figure 7.11. Image of skull and overlying soft tissue rendered transparent to demonstrate soft/hard tissue relationship.

to-back method (the painter's algorithm being a *back-to-front* method). As the name suggests, this involves displaying the objects at the front first, and then keeping a record of which parts of the display screen have been painted so that further attempts to project more distant parts of the object onto these are avoided. In this way it is not necessary to process every surface in the object, as it is in the case of painter's algorithm, so processing time can be considerably reduced. This technique is particularly useful for some types of database structure, and the method has been used for the voxel-based images of the skull presented in this chapter.

In the case of the voxel database, a data structure known as the *octree* has been strongly advocated as being most appropriate to the production of fast computer algorithms for solving the hidden surface problem (Meagher, 1982). The data organization is derived by the successive division of a cube containing the anatomical object into eight equal smaller cubes. If a cube is found to be either empty or full, no further division occurs, but partially filled cubes are further subdivided in the same manner until all elements are either full or empty or some minimum cube size is reached. The advantage cited for this method of data organization is that, from any chosen viewpoint, the visible cubes in a set of eight may be predetermined (Doctor & Torborg, 1981). The hidden surfaces are thus determined at all levels in the octree structure, and the order of display is known without any need of further sorting, which would require much computation and thus takes a long time. At the computational level, the process of determining visibility for the octree is recursive, as was the original subdivision of space. The octree data structure is also very suitable for the implementation of hidden surface algorithms on special purpose hardware.

Once the visible facets or voxel surfaces have been determined, they must now be processed to produce a realistic surface image. In essence this means the calculation of the surface brightness to be applied at each pixel on the screen onto which they are projected, so as to convey to the viewer in so far as possible the three-dimensional character of the anatomy under scrutiny. The simplest method of doing this is by assigning a brightness which decreases in proportion to the distance of the surface element from the viewer. This is known as *depth* or *distance-only* shading, examples of which can be seen in Figures 7.6 and 7.7. Images produced in this way require minimal computation and give an overall impression of depth in an object. Surface details, however, which by their very nature differ only slightly in depth from the surrounding surface, do not show up very well in depth-shaded images. More realistic and detailed images are produced using the physical laws of illumination. With the facet representation, the surface orientation (or *surface normal*) needed for the application of Lambert's law of illumination is immediately available from the vertex coordinates of each facet. When this is combined with the well-established surface-smoothing techniques associated with the names of Gouraud (1971) and Phong (Bui-Tuong, 1975), excellent images are obtained. These methods are well described in stan-

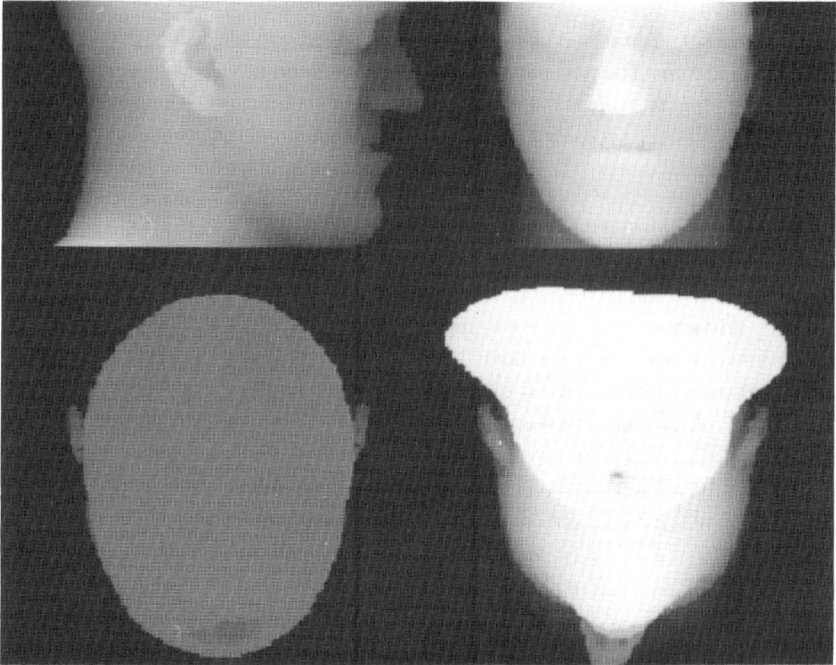

Figure 7.7. Depth shaded images of the face using a voxel database derived from CT scans.

dard texts on computer graphics (Newman & Sproull, 1981). A typical facial image produced by these techniques is shown in Figure 7.8.

For the voxel surfaces, it is more difficult to disguise the discrete structure and to produce good surface rendering. The most recent techniques reported for doing this, although effective, need considerable time for computation (Höhne & Bernstein, 1986; Cline, Lorensen, Ludke, Crawford, & Teeter, 1988). The earliest voxel surface images to be published were based on depth shading and were rather disappointing, lacking as they did in surface detail as well as having a distinct block structure appearance (Herman & Lui, 1979). The databases from which the images were produced were binary arrays, derived by windowing or thresholding the original CT data to achieve speed and have minimum memory requirements. Several techniques which do produce reasonably good images from 3-D binary arrays have now been published. These make use of an estimation of the local surface orientation derived from the relationship between the depths of neighboring visible voxel surfaces (Gordon & Reynolds, 1985). The main limitation of the algorithms using binary data is the resolution with which the surface position is determined, that is to say, in discrete steps of one voxel. An improvement in image quality may be obtained by making use of the original

Hounsfield numbers to determine surface depth at subvoxel resolution and thus a better estimation of surface orientation (Trousset, & Schmidt, 1983; Moss et al., 1987). This method assumes that a surface voxel contains only two components either side of the surface, and uses the Hounsfield number to determine the fraction of each component, and hence the probable location of the surface within the voxel. The normal may then be calculated from the surface depth map as before, but the much-improved resolution leads to more visually satisfying images. This is the technique used to shade the skeletal images presented in this chapter. It has been pointed out that the gradient of the CT data may be used to derive the direction of the surface normal, since at a surface, the direction of the maximum density gradient and the surface normal coincide (Höhne & Bernstein, 1986). Published images using this technique for surface rendering are also of excellent quality. A method has also been reported which derives the surface at subvoxel resolution, and constructs a faceted surface which is then itself shaded

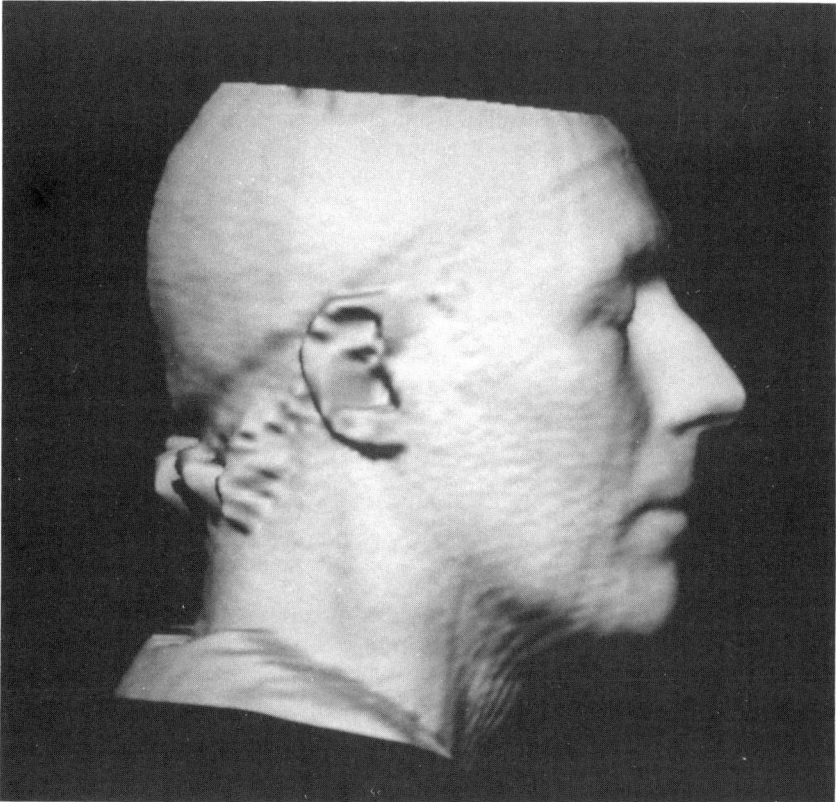

Figure 7.8. Facial image produced by detecting and shading the visible faceted surface.

in the manner described above (Cline et al., 1988). This method in effect allows even the very complex surfaces of facial bones to be facetted.

Simulation of Surgery

A system for planning facial surgery by simulation must be expected to provide a number of essential facilities. The surgeon should be able to interact with the images of the skeletal anatomy, which should be selectable from any viewpoint, so as to rehearse the surgery and anticipate potential problems during the operation. From the surgeon's image manipulations, the system should provide a "prescription" for surgery in terms of osteotomies and the movements of bone fragments required to produce the anatomical rearrangements simulated on the video screen. Very few systems are yet available which meet this requirement. Further, because of the importance of the facial appearance to the patient, it is highly desirable to be able to produce a realistic prediction of the postsurgical facial appearance. Indeed, it may be, as some surgeons have recently proposed, that the goal is set in terms of facial appearance, and the surgery on the facial bones is adjusted to be as compatible as possible with this goal (Kinnebrew et al., 1983). The system should also allow the measurement of dimensions on all parts of the displayed anatomy to be carried out at will. Very few systems meet all of

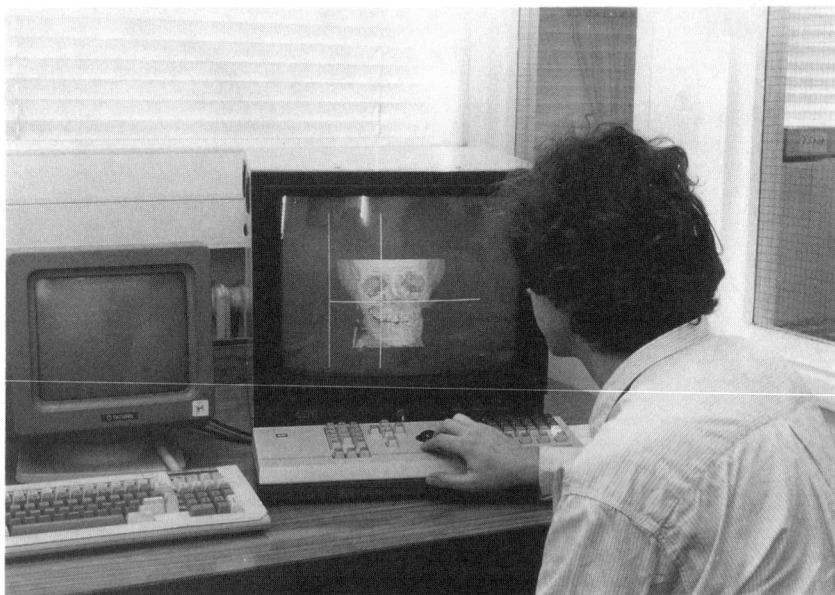

Figure 7.9. General view of the surgeon's workstation showing the trackerball and cursor system and a typical skull image.

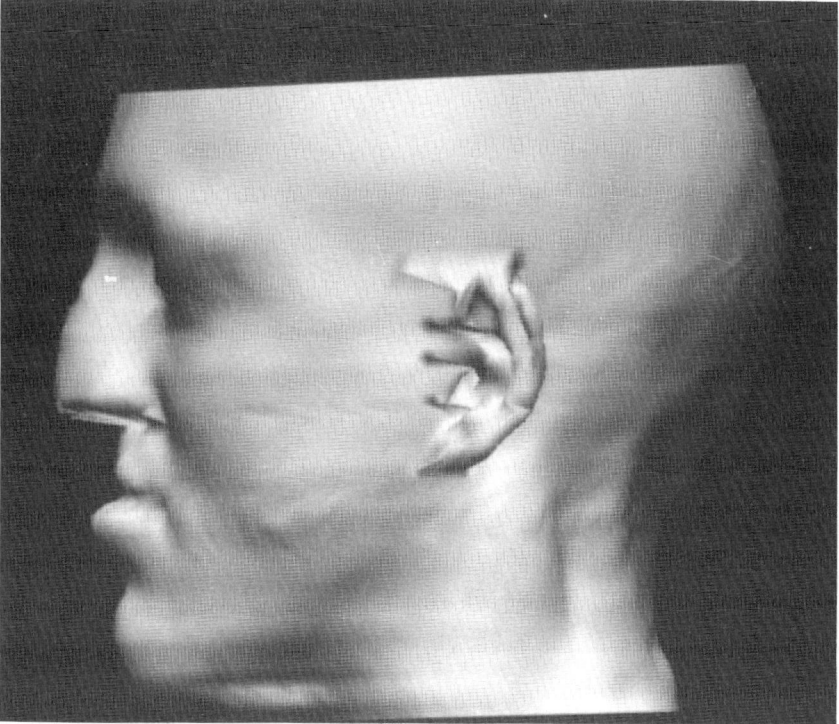

Figure 7.10. Shaded image the facial surface of a patient produced from laser scan data.

these requirements, but the system developed at University College comes close and is the only system so far reported to attempt a prediction of the appearance of the postsurgical face from any viewpoint (Moss et al., 1988).

The simulation of surgical procedures on this system consists of directing and modeling the bone movements under the skin using the database derived from a patient's CT scans. The effects of the bone movements on the facial surface are then calculated, resulting in a procedure specific to the individual needs of each patient.

The osteotomies are almost always planned by displaying an image of either a lateral or anterior view of the skull on a video monitor as shown in Figure 7.9. An image of the face may also be displayed to assist assessment (Figure 7.10). The soft tissue may also be displayed as a translucent layer over the bones, so that the relationship between the face and its supporting bone structure may be fully appreciated (Figure 7.11). An area is then delineated by using a trackerball-driven cursor to mark individual points around its boundary. These points are then connected by the computer to form a closed polygon (Figure 7.12). The

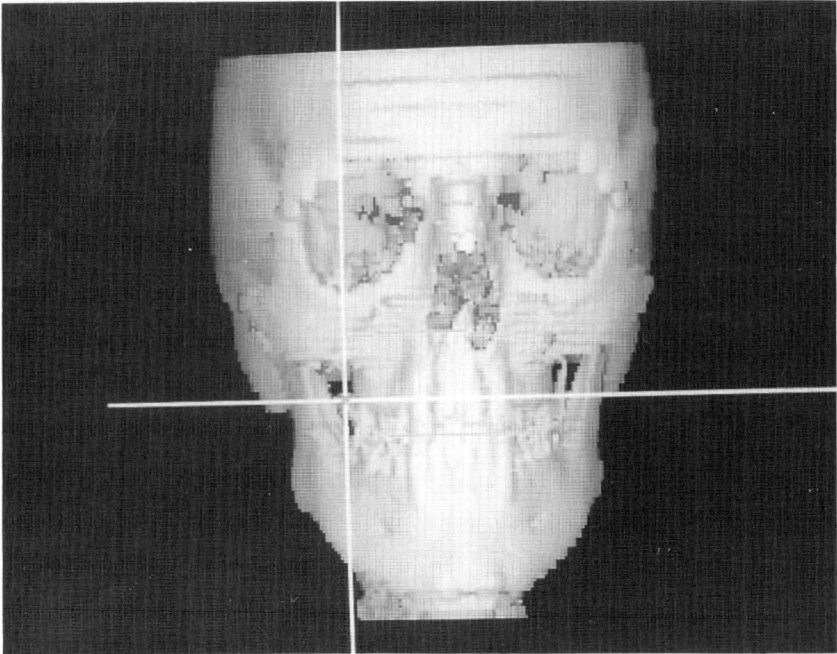

Figure 7.12. Marking an outline to sweep out a volume of interest.

volume of the anatomy selected by this procedure is the volume which would be swept out by the bounded area moving through the skull perpendicular to the surface of the screen. This has been likened to the cutting of pastry shapes. The volume of interest so defined is now displayed separately on another part of the screen and, if desired, from another view. It may then be further cut into smaller pieces by the same procedure, as shown in Figure 7.13. In the case illustrated the second stage is performed to remove the vertebrae which have been included in the swept volume. Volumes of the anatomy isolated in this manner may now be moved to a new position with respect to the remainder of the face and skull by marking the direction of movement required in two orthogonal views. The repositioned bone fragments are then displayed in the new relationship with the unchanged parts of skull (Figure 7.14), and the result may be viewed from any chosen direction, so that the new configuration of the bony anatomy may be assessed. This procedure is repeated as often as necessary to produce a satisfactory result.

The same procedure which is used to simulate, and hence plan, the surgery may naturally be used at an early stage for isolating or uncovering elements of the anatomy for closer scrutiny so as to improve or confirm initial diagnoses.

From the arrangement of the bony anatomy a prediction of the postsurgical facial appearance may now be developed. Initially in the database, the soft tissue has simply been moved by the same amount and in the same direction as the underlying bone. By selecting the threshold range appropriate for soft tissues, this crude estimation of facial shape may be displayed (Figure 7.15). Since, however, no modeling of the skin surface has taken place, and the soft tissues have been permitted to follow underlying bone movements, large steps appear in the skin surface. To generate smoother images corresponding to the actual behavior of soft tissue as it accommodates to the supporting bone requires that the modified skin surface remains smooth and continuous. To achieve this, 2-D axial sections from the modified database lying parallel to the original CT sections are generated and displayed. The contours of the skin surface are then found for each section by simple contour following. These are smoothed out by a very simple model which will undoubtedly need refinement when enough data exist. In this model, we imagine the soft tissues to be anchored to a point remote from the bone movement (e.g., near the ears), and then linearly interpolate skin surface displacement between this point and the step produced in the surface in response to bone movement. A typical result for a prediction of post surgical facial shape

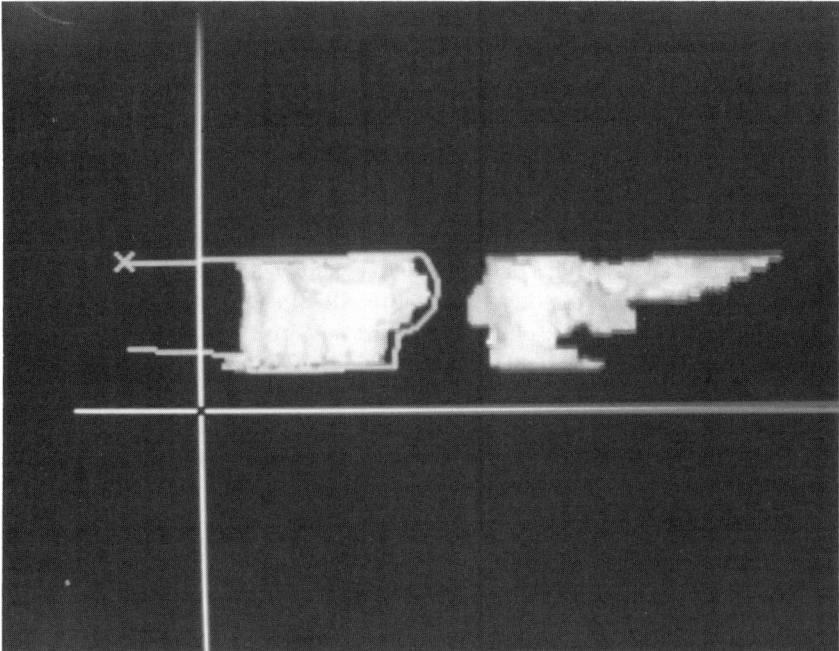

Figure 7.13. Further resectioning of the volume of interest.

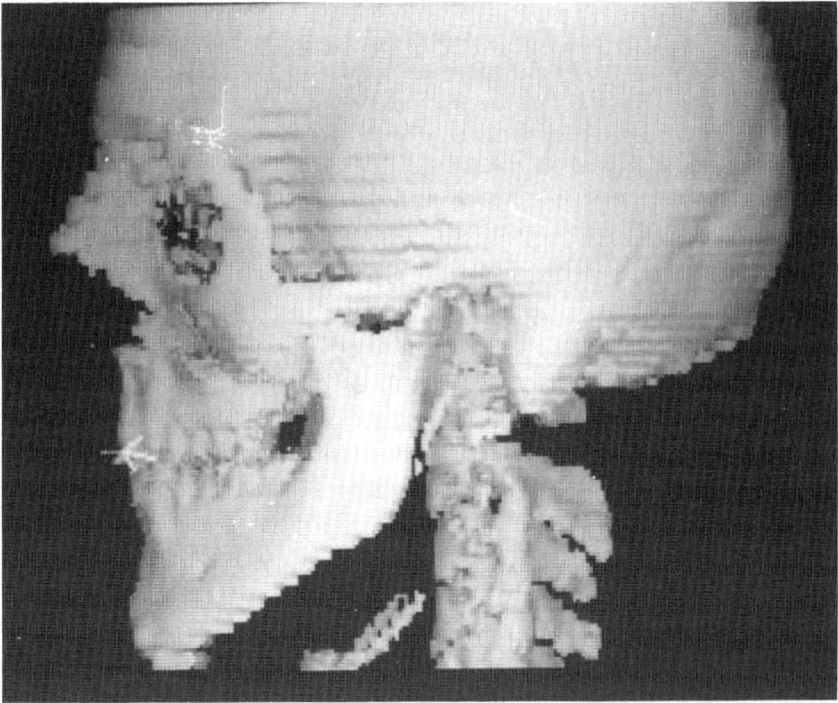

Figure 7.14. Applying the bone fragment movement required.

is shown in Figure 7.16. This model is clearly only intended as a first approximation, and is necessarily limited, since only small scale studies of soft tissue
response to facial osteotomies in the midline (Freihofer, 1977; Dann, Fonseca, &
Bell, 1976; Willmot, 1981; Suckiel & Kohn, 1978) and on a small number of
other vertical profiles (Burke et al., 1983) have so far been reported. The patient
protocol which has been established for using the University College System
includes the laser scanning of the patient's face at regular intervals after surgery,
together with a single limited set of CT scans. These measurements will be used
to develop and refine the soft tissue modeling by extending the information on
soft tissue behavior to most of the facial surface.

The final stage is to produce information in a form which can be used by the
surgeons in the operating theater. This is currently provided in the form of
pictures produced (e.g., Figure 7.17) by a video printer and a list of osteotomies
and bone movements in millimeters. The system can also generate profiles of the
newly modeled skeletal surface from which fixing templates (Champy plates)
could be manufactured. This last capability, however, has not yet been used in
actual surgery.

Other Facilities

A number of supporting features have been found to be important to facilitate the planning and analysis of facial surgery and the monitoring of facial growth, which is often required in order to carry out the surgery at the best possible time to produce a satisfactory long-term outcome.

The numerical analysis of orthognathic surgery is usually expressed in terms of a changing relationship between a number of anatomical landmarks, and the planes and angles constructed from them (Epker & Fish, 1986). The analysis has always at some stage been presented in some graphic form. A means of marking and recording the positions of landmarks is therefore essential. Using the graphics facilities provided, the operator is able to choose a viewpoint and direction of illumination which allows the required landmark to be most easily seen. The landmark is selected by using a trackerball to place the screen cursor over it. The system computes the three-dimensional coordinates of marked points and the distances between them. Figure 7.18 illustrates how landmarks are typically marked after selection. The results obtained using this technique have been found to be highly reproducible.

The presentation of profiles has also been traditionally used to represent the

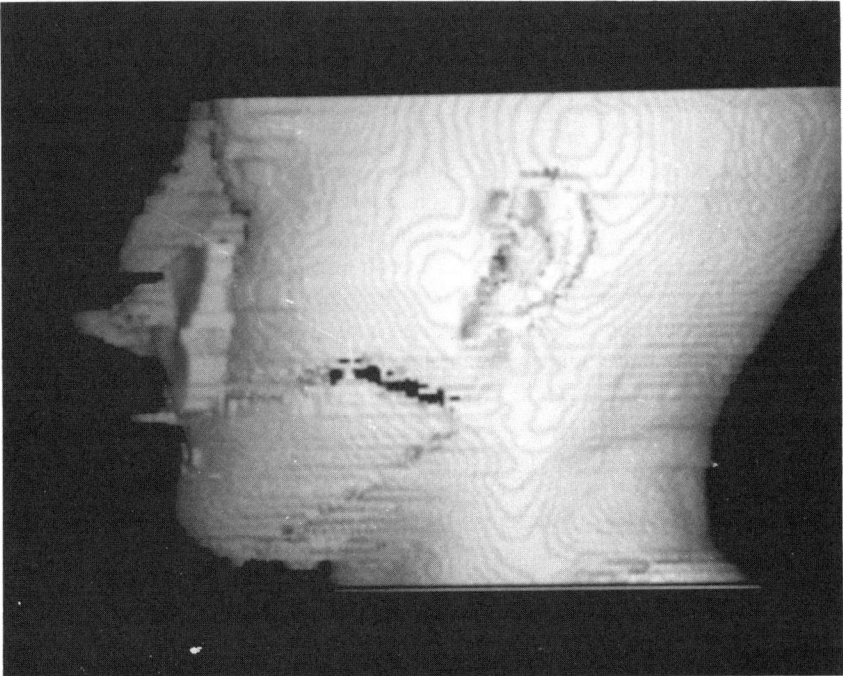

Figure 7.15. First approximation to facial surface.

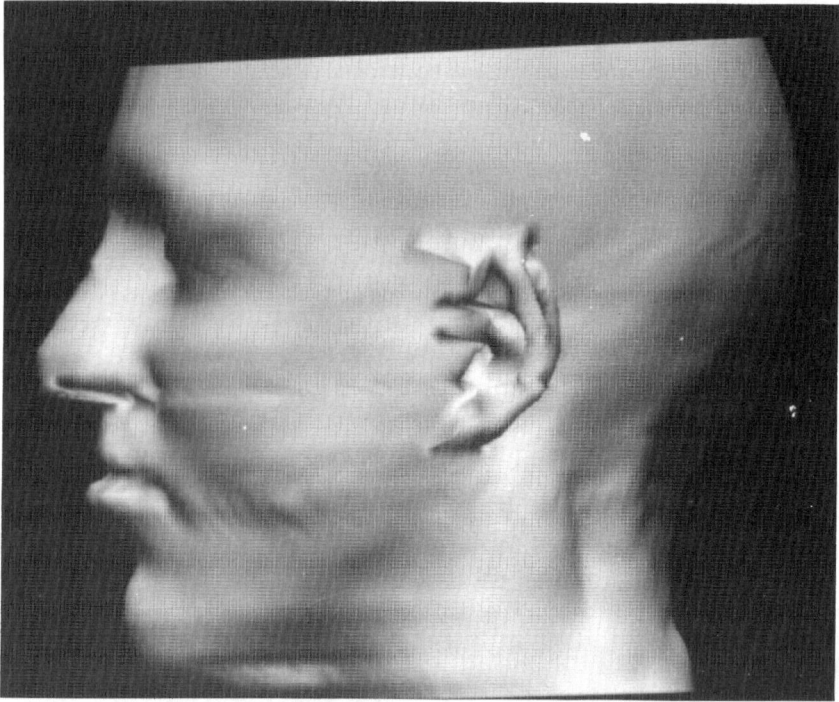

Figure 7.16. Prediction of the postsurgical facial shape.

effects of facial surgery along the midline (Epker & Fish, 1986). The current system will generate the profile between any two points on both the facial and skeletal surfaces. The user chooses a suitable viewpoint, and marks two points on the surface as described above. The profile is generated by computing the points at which a plane perpendicular to the image plane containing the two marked points cuts the 3-D surface. An example is shown in Figure 7.19. The intervals at which these points are calculated is controlled by the user. Up to 256 points per profile are computed. Profiles can be plotted automatically on an X-Y plotter as shown in Figure 7.20, or stored for future use.

The measuring techniques have been validated using a dry skull and a Roman bust as subjects which were CT and laser scanned, respectively. Distances measured by marking points on the images and by the use of calipers were found to be in good agreement, with differences of usually less than 1 mm being recorded.

The other important requirement for the comparison of anatomical surfaces is the ability to register them in some way. To achieve this, the two surfaces to be compared, which might, for example, be the pre- and postoperative face from laser scan data, are displayed side by side on the screen. An area which has not been altered by surgery is now identified. This will usually be the forehead. The relative orientation of the two surfaces is determined by marking three corresponding points on each surface from which a transformation matrix to bring the two surfaces into the same position and orientation is calculated. Several options have been provided to make this task easier. In one case, the midline only need be marked (as for selecting profiles), and two points located on the forehead at set distances either side of the midline are automatically computed. The registered images may be superimposed from any viewpoint to display differences, or an image representing the distances between the two surfaces along a selected viewing line can be displayed. This facility, as well as being essential for the study of the effects of surgery, has proved useful in growth studies and the monitoring of facial tumors.

Discussion

To establish its effectiveness, the system at University College is now under evaluation, and has been used clinically in the planning of some 20 cases to date

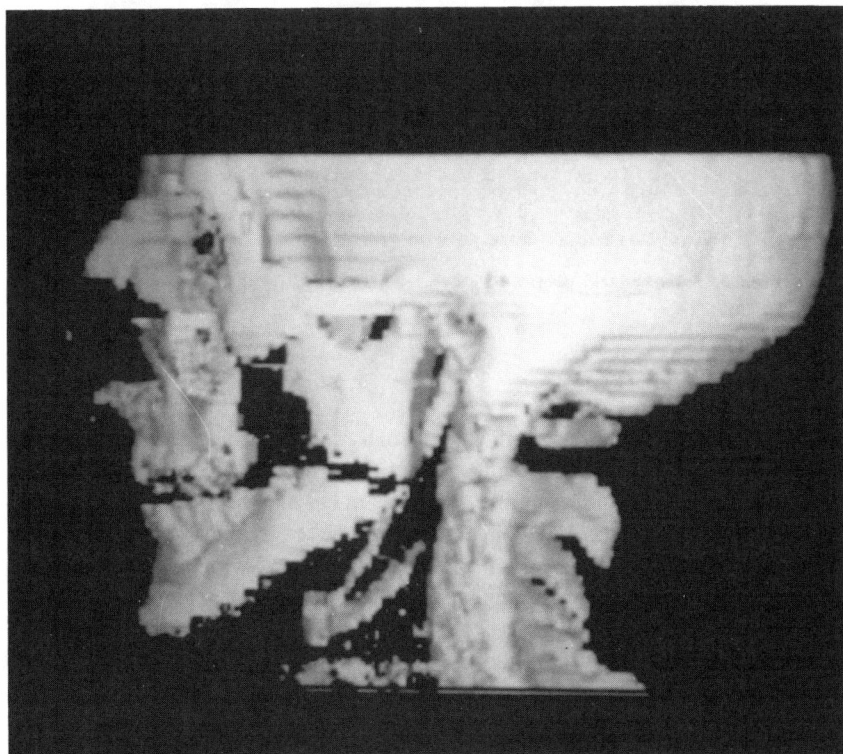

Figure 7.17. Image typical of a more complicated bimaxillary operation.

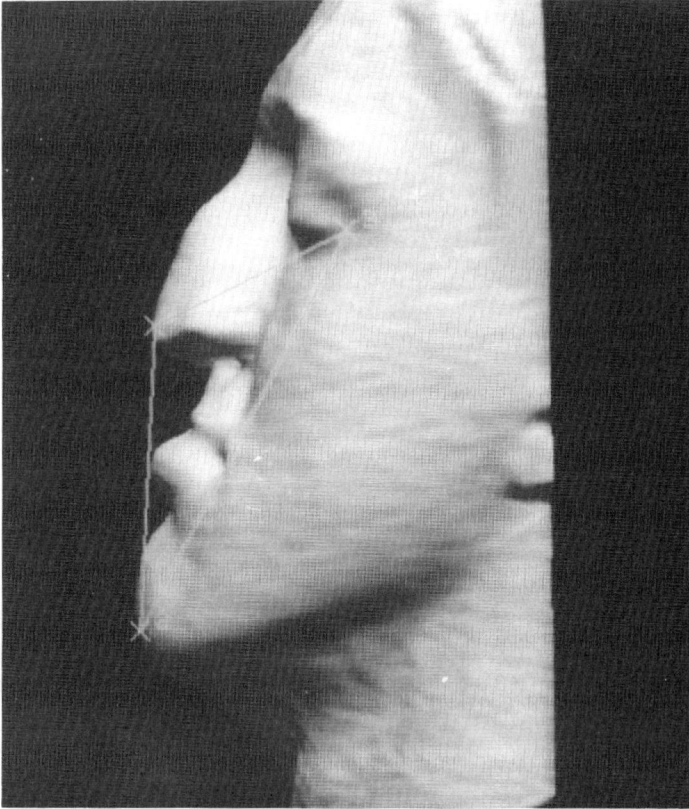

Figure 7.18. Marking points for 3-D measurements on the facial surface.

(1988). The types of surgery performed usually involve bimaxillary procedures of the Le Fort and Kufner types, to correct for differential development in the growth of the maxilla and mandible. The system has also been used for diagnostic purposes in many cases of congenital abnormality, such as cleft palate, and facial deformity where the skull is so asymmetrical that, without the aid of 3-D reconstructions, the surgeon is unable to preplan the surgery until the time of incision. The ability to visualize the 3-D skull has already proved to be an invaluable tool to both cranio- and maxillofacial surgeons.

In order to aid the planning of maxillofacial surgery, it has recently been pointed out that a 3-D normative standard for the face and the skull, together with an analytical path from an abnormal condition to this standard, are needed

(Cutting et al., 1986). Such standards are commonly used in 2-D planning, where systems of analysis have developed as a result of experience over many years (Jacobson & Cranfield, 1985). The establishment of a normative standard, and the quantitative evaluation of the use of graphics in this field, require both sufficient data and methods for the mathematical description and analysis of the facial form in three dimensions. The analysis of shape and changing form in two dimensions has received much attention in the literature and in orthodontics manuals (Jacobson & Cranfield, 1985) and some potentially useful 3-D methods using surface curvature have recently been described (Coombes, Linney, Grindrod, Moss, & Moss, 1988). Now that it is possible with the type of system described here to produce accurate measurements distributed over the whole facial surface, it will be possible to test how well these techniques can be applied to this problem and to come to a clearer understanding of facial aesthetics.

The systematic production of anatomical models, fixing plates and prostheses to replace or augment the craniofacial bones is a potentially important area for

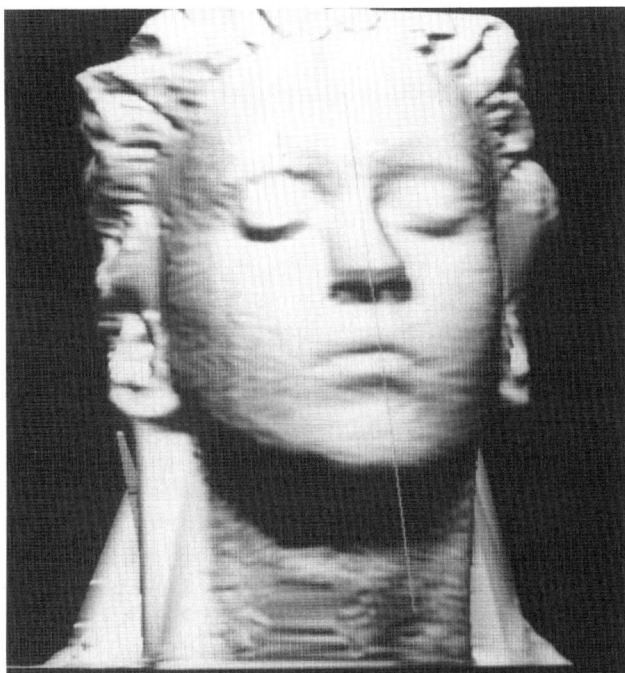

Figure 7.19. Profile marking on facial image.

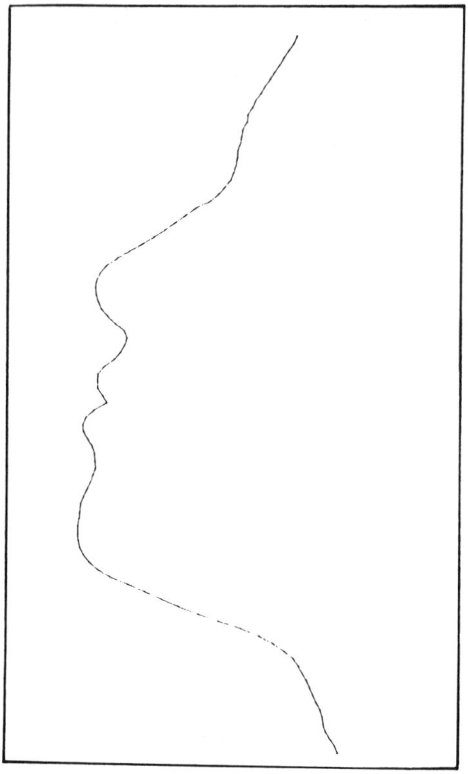

Figure 7.20. Plot of the marked profile shown in Figure 7.19.

development. Some systems have already been described for model production and the manufacture of a large range of prostheses (Rhodes, Kuo, & Rothman, 1987). For these facilities to be come into wider use, it will be necessary to develop programs for the numerical control of milling machines using data generated by surgical planning packages. The establishment of a suitable operating protocol would clearly be a major step forward in this direction.

The main improvement which is sought after in new systems is that of increased speed of response. This is of the order of minutes for some procedures on the system described here, and clearly a response time of a few seconds or less would be highly desirable. It is expected that work currently underway will produce a 3-D medical graphics work station based on Transputer technology which will have response times of this order (Tan, Richards, & Linney, 1988).

There is still an active controversy on the question of how accurate the 3-D presentations are, and especially of how serious is the presence of pseudo-foramina in terms of their possible effect on clinical diagnosis (Tessier & Hemmy, 1985, 1986). This will no doubt be resolved in time, as more clinicians use

3-D medical graphics techniques to study an ever widening range of diseases, injuries, and deformities.

FUTURE DEVELOPMENTS: COMPUTER
AND DISPLAY SYSTEMS

Many of the early systems for the display of the three-dimensional human anatomy were disappointing, in part because of the time taken to produce an image and also because of low image quality. Some early papers were pessimistic both about the speed and also the cost of using graphics techniques either for diagnostics or for planning surgery (Savolini, Cabanis, Iba-Zizew, DeNicola, & Hemmy, 1984).

Since that time the resolution of display monitors has improved, and so has the speed of computation. Graphics techniques, especially those for surface rendering of the voxel arrays, have also improved, giving rise to far better quality images.

Clearly important features of any system are speed and a high-quality display system. Advancing technology has now produced a wide variety of adequate display devices, but there are still two schools of thought concerning speed. On the one hand, there are those who have proposed and developed special purpose hardware for performing the graphics algorithms (Goldwasser et al., 1988) and on the other hand there are those who are attempting to exploit general purpose computing systems (Tan et al., 1988). The special purpose hardware is generally very fast, but is usually constructed around a particular kind of database to which the user is then restricted. The development of new programs and routines on such hardware is often something which only the manufacturer can do, imposing further restrictions on the user. Special purpose hardware design has also been largely directed towards the problem of display and not towards the interactive alteration of three-dimensional anatomical images required for the simulation of surgical procedures.

Although general purpose computing systems are slower at producing 3-D images than special purpose hardware at the same cost, their speed is gradually improving. The introduction of general purpose systems designed for parallel processing also now offers new opportunities for producing images at satisfactory speed. The parallel computing element known as the Transputer may prove to be an important device in this respect. The big advantage these general purpose modular systems have, is that they can be programmed to the changing requirements of individual clinics, and new concepts of diagnosis and surgical planning can be incorporated as they arise.

There are now a number of commercially available systems for the production of 3-D images from CT scanners, but most do not offer the possibility of simulating surgery. These systems have recently been reviewed by Goldwasser et al. (1988).

CONCLUSIONS

A system has been developed which uses computer graphics for the simulation, planning, and prediction of maxillofacial surgery. The data used are provided by a custom-built facial surface laser scanner and standard X-Ray CT medical imaging systems. The system is currently in clinical use and undergoing evaluation. Mathematical techniques are being developed for the necessary case analyses.

REFERENCES

Arridge, S., Moss, J.P., Linney, A.D., & James, D. (1985). Three dimensional digitisation of the face and skull. *Journal of Maxillofacial Surgery, 13*, 136–143.

Barrett, C.N. (1984, March). Preoperative planning with interactive graphics. The dawning of a new era. *Computer Graphics World*, pp. 11–18.

Bhatia, S. N., & Sowray, J.S. (1984). A computer aided design for orthognathic surgery. *British Journal of Oral & Maxillofacial Surgery, 22(4)*, 237–253.

Brewster, L.J., Trivedi, S.S., Tuy, H.K., & Udupa, J.K. (1984, March). Interactive surgical planning. *IEEE Computer Graphics and Applications, 4*, 31–40.

Bui-Tuong, P. (1975). Illumination for computer generated pictures. *Communications of the Association of Computing Machinery (ACM), 18*, 311–317.

Burke, P.H., Banks, P., Beard, L.F.H., Tee, J.E., & Hughs, C. (1983). Stereophotographic measurement of change in facial soft tissue morphology following surgery. *British Journal of Oral Surgery, 21*, 237–245.

Cline, H.E., Lorensen, W.E., Ludke, S., Crawford, C.R., & Teeter, B.C. (1988). Two algorithms for the three-dimensional reconstruction of tomograms. *Medical Physics, 15*, 329.

Coombes, A.M., Linney, A.D., Grindrod. S.R., Mosse, C.A., & Moss, J.P. (1988). 3-D measurement of the face for the simulation of facial surgery. *Proceedings of the 5th. International Symposium on Surface Topography and Body Deformity*. Stuttgart & New York: Gustav Fischer Verlag.

Cutting, C., Grayson, B., Bookstein, F., Fellingham, L., & McCarthy, J.G. (1986). Computer-aided planning and evaluation of facial and orthognathic surgery. *Computers in Plastic Surgery, 13*, 449–462.

Dann, J.J., Fonseca, R.J., & Bell, W.H. (1976). Soft tissue changes associated with total maxillary advancement: A preliminary study. *Journal of Oral Surgery, 34*, 19–23.

Doctor, L.J., & Torborg, J.G. (1981). Display techniques for octree encoded objects. *IEEE Computer Graphics and Applications, 1*, 29–38.

Epker, B.N., & Fish, L.C. (1986). *Dentofacial deformities. integrated orthodontic and surgical correction* (Vol. 1). St. Louis, MO: C.V. Mosbey.

Flynn, M., Matteson, R., Dickie, D., Keyes, J.W., & Bookstein, F. (1983). Requirements for the display and analysis of three-dimensional medical image data. *SPIE Picture Archiving and Communication Systems for Medical Applications, 418*, 213–224.

Freihofer, H.P. (1977). Changes in nasal profile after maxillary advancement in cleft non-cleft patients. *Journal of Maxillofacial Surgery, 5*, 20–27.

Gholkar, A., Gillespie, J.E., Hart, C.W., Mott, D., & Isherwood, I. (1988). Dynamic low-dose three dimensional computed tomography: A preliminary study. *British Journal of Radiology, 61*, 1095–1099.

Goldwasser, S.M., Reynolds, R.A., Talton. D.A., & Walsh, E.S. (1988). Techniques for the rapid display and manipulation of 3-D biomedical data. *Computerized Medical Imaging and Graphics, 12*, 1–24.

Gordon, D., & Reynolds, R.A. (1985). Image shading of 3-dimensional objects. *Computer Vision, Graphics and Image Processing, 29*, 361–376.

Gouraud, H. (1971). Continuous shading of curved surfaces. *IEEE Transactions on Computers, C–20(6)*, 623–628.

Hall, D.H. (1985). Plannning techniques. In W.H. Bell (Ed.), *Surgical correction of dentofacial deformities—new concepts* (pp. 153–160). Philadelphia, PA: W.B. Saunders.

Herman, G.T., & Liu, H.K. (1979). Three-dimensional display of human organs from computer tomograms. *Computer Graphics and Image Processing, 9*, 1–21.

Höhne, K.H., & Berstein, R. (1986). Shading 3-D images from CT using gray-level gradients. *IEEE Transactions on Medical Imaging, 5*, 45–48.

Hounsfield, G.N., Ambrose, J., Perry, J., & Bridges, C. (1973). Computerised transverse axial scanning (tomography). *British Journal of Radiology, 46*, 1016–1051.

Jacobson, A., & Cranfield, P.W. (1985). *Introduction to radiographic cephalometry*. Philadelphia, PA: Lea and Febiger.

Kinnebrew, M.C., Hoffman, D.R., & Carlton, D.M. (1983). Projecting the soft-tissue outcome of surgical and orthodontic manipulation of the maxillofacial skeleton. *American Journal of Orthodontics, 84*, 508–519.

Meagher, D.J. (1982). Geometric modelling using octree encoding. *Computer Graphics and Image Processing, 19*, 129–147.

Moss, J,P., Grindrod, S.R., Linney, A.D., Arridge, S.R., & James, D. (1988). A computer system for the interactive planning and prediction of maxillo-facial surgery. *American Journal of Orthodontics and Dentofacial Orthopaedics, 94*, 469–475.

Moss, J.P., Linney, A.D., Grindrod, S.R., Arridge, S.R., & Clifton, J.S. (1987). Three-dimensional visualization of the face and skull using computerized tomography and laser scanning techniques. *European Journal of Orthodontics, 9*, 247–253.

Newman, W.M., & Sproull, R.E. (1981). *Principles of interactive computer graphics*. New York: McGraw Hill.

Norwood, H.M., Cunningham, C., Bowesley, S., & Taylor, C. (1988). Patient absorbed dose for the Philips Tomoscan 350 CT scanner: A repeat study. *British Journal of Radiology, 61*, 639–640.

Rabey, G.P. (1977). Current principles of morphanalysis and their implications in oral surgery practise. *British Journal of Surgery, 15*, 97–134.

Ramachandran, V.S. (1988). Perception of shape from shading. *Nature, 331*, 133.

Rhodes, M.L., Kuo, Y.M., & Rothman, S.L.G. (1987). An application of computer graphics and networks to anatomic model and prosthesis manufacturing. *IEEE Computer Graphics and Applications, 7*, 12–25.

Savolini, U., Cabanis, E.A., Iba-Zizen, M.T., De Nicola, M., & Hemmy, D.C. (1984).

Apport diagnostique de la reconstruction tri-dimensionelle en scanner RX: Coupes et surfaces de l'anatomie cephalique. *Annales de Chirorgie Plastique et Esthetique, 29(4)*, 339–357.

Suckiel, J.M., & Kohn, W,M. (1978). Soft tissue changes related to the surgical management of mandibular prognathism. *American Journal of Orthodontics, 73*, 676–680.

Tan, A.C., Richards, R., & Linney, A.D. (1988). 3-D medical graphics—using the T800 transputer. Developments in using OCCAM. OUG–8. *Proceedings of the 8th OCCAM User Group Technical Meeting*, IOS, Amsterdam.

Tessier, P., & Hemmy, D. (1985). CT of dry skulls with craniofacial deformities: Accuracy of three-dimensional reconstruction. *Radiology, 157,* 113–116.

Tessier, P., & Hemmy, D. (1986). Three-dimensional imaging in medicine—A critique by surgeons. *Scandinavian Journal of Plastic Reconstructive Surgery, 20*, 3–11.

Trousset, Y., & Schmitt, F. (1983). Active ray tracing for 3-D medical imaging. In G. Marechal (Ed.), *Eurographics '87*. Amsterdam: Elsevier.

Vannier, M.W., Marsh, J.L., & Warren, J.O. (1983). Three-dimensional computer graphics for cranio-facial surgical planning and evaluation. *Computer Graphics, 17(3)*, 263–273.

Willmot, D.R. (1981). Soft tissue profile changes following the correction of Class III Malocclusions. *British Journal of Orthodontics, 8*, 175–182.

8
Animation using Image Samples

Neil D. Duffy
Department of Electrical and Electronic Engineering
Heriot-Watt University
Edinburgh, Scotland

INTRODUCTION

Methods of creating synthetic images of 3-D scenes fall into one of two general categories. The first uses a direct digital representation of the surface to be modelled, storing either a complete set of surface points or, more usually, lists of the edges and vertex positions of an underlying polygon model. The second method uses a parametric description of the surface.

Animation using the second technique is useful where some creative input is needed to generate the image; a method described by Nahus, Huitric, and Saintoureus (1988) uses B-Splines. Parametric methods are less useful if the image to be created is of some existing 3-D scene. This is particularly true if the image to be created is to be of a recognizable person.

Using the first method referred to above, and assuming that digitized surface information relating to the underlying 3-D image is available, the conventional technique for reconstructing an image is to use wire frame graphics techniques in which animation is achieved by manipulation of the x,y, and z coordinates of the underlying model. To achieve realism, the visible surfaces of the wire frame model are identified, and the color and intensity of the surface are computed using a shading model. Facial animation using this approach was first described by Parke (1972), who carried out surface measurements of a subject's face manually. The face was then animated by applying 3-D transformations to the vertices of the underlying model and shading the polygon facets appropriately. Features within the face, such as eyebrows, eyeballs, mouth, and lips, were modeled separately.

In terms of realism, the shortcomings of the shaded/wire frame technique are that it produces images which have a rather synthetic visual quality. This is because shading models operate by calculating the light that will be reflected from a smooth surface when it is illuminated by some known light source. They have no knowledge of, and are unable to portray, fine surface detail such as facial lines and wrinkles, skin texture, moles, and freckles.

For animation purposes, images based on the use of the shaded/wire frame technique can be manipulated to produce convincing results. Groups of polygon vertices relating to muscle models can be manipulated to produce particular facial expressions such as smiles or frowns, as described by Waters (1987). Again the results of the animation tend to have a synthetic quality which is ideal for the creation of cartoon or caricature type images. The difficulty in applying realistic, life-like animation to such images lies in the problem of providing sufficient model surface detail to model accurately features such as the eyes or mouth. In the latter case this can be illustrated by considering the detail required to model the mouth for situations where animated speech is required. Various mouth positions require various amounts of tooth exposure. This in turn requires an underlying model of part of the mouth interior which is exposed appropriately during the speech act. Parke (1982) uses such a method, in which he defines mouth expressions in two dimensions and interpolates between the extremes of open and closed. The resulting two-dimensional vertex positions are then projected onto an underlying surface. This technique still results in cartoon-like animation of facial features because of the difficulty in obtaining underlying models of sufficient likeness to real subjects.

The shaded wire frame technique is ideal for many purposes but there are some applications, particularly in the area of facial animation, where more realism is required. These are discussed briefly in the "Applications" section, and include visual telecommunications, where an image of a speaker is to be reproduced at the far end of a communication channel which has insufficient bandwidth to transmit the picture in the normal manner; and forensic identification, where a witness may have seen a suspect from a viewpoint other than that held on photographic record.

For applications of this kind, where highly realistic reconstruction and animation of facial images is required, an alternative approach can be used whereby a two-dimensional picture of the facial surface detail (the base image) is mapped on to an underlying wire frame model using a texture mapping technique. Gross animation involving changes in head size, position and orientation can be achieved by conventional coordinate transformation techniques. Facial animation of eye and mouth movements can be achieved by blending sequences of sub-images onto the base image. The texture-mapping technique is an extension of the conventional wire frame/shading method, and achieves the appearance of modeling complex surface detail in an efficient manner without the requirement to model the actual fine detail of the surface. The technique can be combined with conventional shading methods to provide additional realism by the addition of shadows caused by the position of synthesised light sources, and can be combined with muscle model techniques to provide a wide range of facial expressions.

THE UNDERLYING 3-D MODEL

If the reconstruction process is to achieve realistic results, it is vital that the underlying 3-D model accurately represents the shape of the facial surface and that there is accurate registration between the model and the base image. The result of using inaccurate models, such as cylinders, to map the base image onto is that only limited rotational transformations of the image can be made before the realism of the reconstructed image is lost.

Derivation of Surface Information

The traditional way to obtain 3-D surface information is to take manual measurements of vertex positions which have been marked on a subject's face, or on the surface of a plaster model of a face. This is a time-consuming activity that is tedious if a number of subjects have to be modeled.

One method which avoids the need for repetitive measurements on a number of subjects is to create a universal 3-D surface model and to adapt it to fit the subject being modelled, firstly so that the height and width of the model match the base image dimensions, and second to match the model's feature positions, such as the mouth, nose, and eyes, to the feature positions of the base image. Points on the contour of the surface model are then adjusted so that they coincide with the contours of the subject's face. This method has been adopted by Aizaw, Yamada, Harashima, and Saito (1987), but still requires some 3-D measurements of the subject's face to be carried out.

3-D measurements of facial surface detail can be carried out automatically using structured lighting methods. Using these methods, a line of light is produced by a projector or created using a low power laser whose beam is passed through a cylindrical lens. The line is projected onto the subject's face at a known position and illuminates a facial contour. By using a TV camera and viewing the illuminated contour at an angle, precise measurements of the profile of the contour can be made. To build up a complete 3-D image, successive contours can be illuminated by either projecting a vertical line in a fixed position and rotating the subject around the vertical axis, or by holding the subject in a fixed position and altering the position of the projected line by the use of a mirror system as illustrated in Figure 8.1. The latter method is quicker (5 seconds to capture an image) and cheaper but only allows modeling of frontal facial surface detail. The illustrations used in this chapter have all been produced using a 3-D model derived by this process.

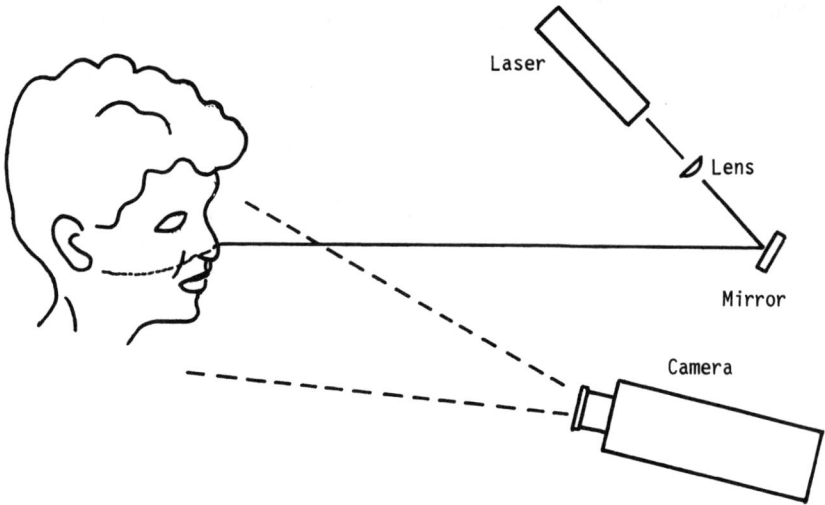

Figure 8.1. Scanning apparatus for 3-D data capture.

Data Reduction

Automatic methods of acquiring 3-D image data yield a large number of data points which need to be to be reduced to a manageable number in order to generate a 3-D polygon model. The scanning process shown in Figure 8.1 typically yields 256*256 points. In practice it is found that a sufficiently detailed polygon model can be derived from the storage of 400 vertex points. Selecting 400 vertex points from the original 65,536 so that they form a symmetrical mesh in the xy plane is a simple operation and results in a surface model such as that shown in Figure 8.2a. In practice it has been found that a more efficient model in terms of the texture mapping process is obtained when the polygons are formed in such a way that small polygons are created covering the mouth, eye, and nose areas, where the surface details are complex, and large polygons are used where the surface is featureless. The surface shown in Figure 8.2b has a quarter of the number of vertices of that of Figure 8.2a.

Boissonnat (1984) has described a volume-based algorithm which will reduce the data points in execution time o(N^2 log N) where N is the number of surface points. A faster algorithm executing in time proportional to N has been described by Duffy and Yau (1988). This latter algorithm is surface based and is ideally suited to reducing data from a scanning system such as that described above. Polygons in the form of triangles are fitted to the surface in such a way that the plane of each triangle is tangential to the surface. The basis of this method is illustrated in Figure 8.3a in which V_1 and V_2 are vertices which have already been identified by the triangulation process. Original data points which lie in the direction of bisector MB of V_1V_2 are selected. These data points are considered in turn as candidates for vertex V_3, the vertex required to complete the triangle V_1, V_2, V_3. The segments MA_1, MA_2, MA_3 . . . MA_N are in the plane of the prospective triangle. In Figure 8.3b, A_4 is being considered as a candidate for vertex V_3. The perpendicular distances δ_1, δ_2 and δ_3 between data points A_1, A_2 and A_3 and segment MA_4 are calculated. A_4 is selected as a vertex provided that none of δ_1, δ_2 or δ_3 exceed a limit value δ_{lim}. In practice further tests are carried out before A_4 is accepted as a vertex. These tests are to ensure that δ_{lim} is not exceeded along the edges V_1V_3 and V_2V_3 of the new triangle, and to ensure that triangles are generated evenly without being excessively acute. The size of the limit value determines the size of the polygons created for a given surface curvature and thus the number of polygons created for the complete 3-D model.

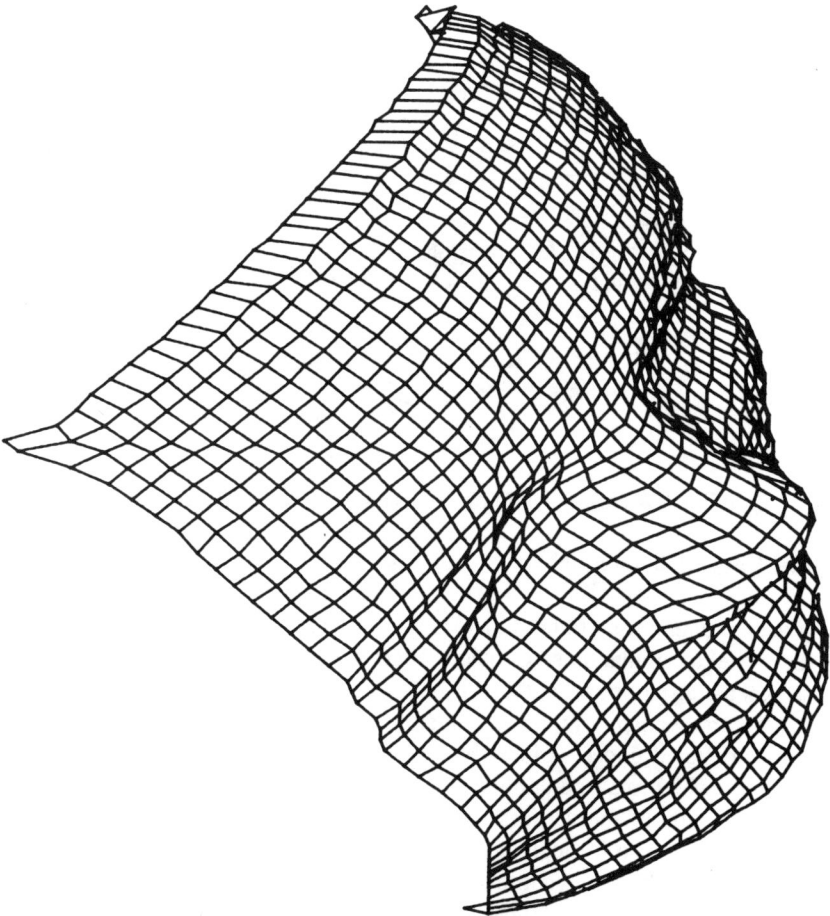

Figure 8.2(a). Polygon model using a symmetrical mesh.

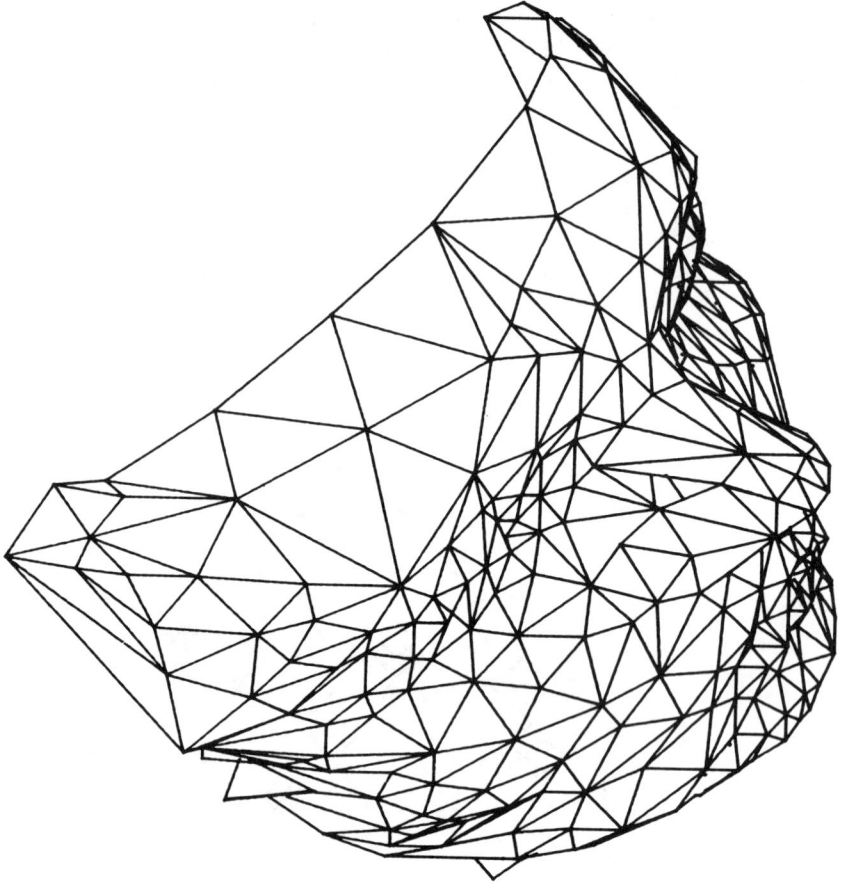

Figure 8.2(b). Polygon model using variable sized triangles.

Model Data Structure

Any data structure used to hold the polygon model must allow the manipulation of the model as animation takes place. There are two observations which can be made when considering what type of data structure is most appropriate. The first is that the arrangement of the network structure of the model, i.e., the topology, remains unaltered under various transformations, such as scaling, rotation, or translation, which may be applied to animate it. During these transformations only the position of the vertices is redefined. One suitable arrangement which takes advantage of the topological invariance of the model is a hierarchichal arrangement in which polygon data are stored in two lists.

The first, an edge list, holds an edge reference and references to the two

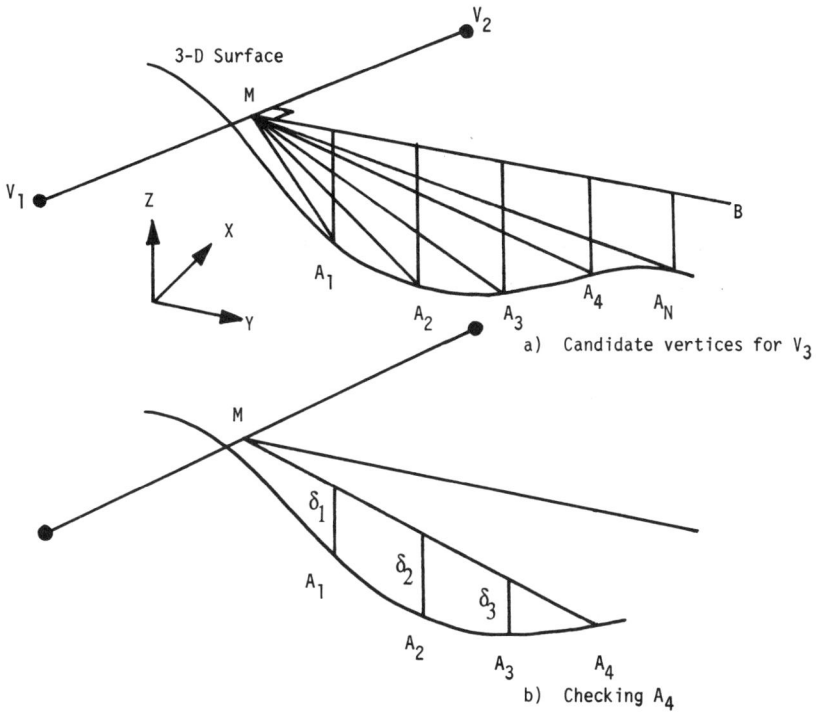

Figure 8.3. Automatic triangulation process. (a): Candidate vertices for V3. (b): Checking A4.

associated vertices; the second is a list of vertices in terms of their xyz cartesian coordinates. With this arrangement of data storage, coordinate transformations that take place during the animation process only affect the data in the vertex list.

BASE AND SUBIMAGES

A property of the texture mapping technique described later (see "Generating the Synthetic Image by Texture Mapping") is that the image which is mapped on to the 3-D model is a frontal view of the face and is invariant to any scaling, translation, or rotation transforms which may be applied to animate the reconstruction. This image is called the base image. Animation of facial detail such as eye movement or speech is achieved by blending sequences of eye and mouth subimages into the base image at appropriate positions. The subimages are also stored as frontal views, invariant to transforms applied during the animation process.

Derivation of the Subimages

At present the best method for obtaining a sequence of subimages is to extract the images manually from a video of the original subject. To simplify the extraction of the subimages from the frames, it is best to ensure that their positions are relatively fixed. This can be achieved by recording the set of frames containing the subimages over a short time period and in a single sitting.

Some success has been achieved by Yau and Duffy (1989) in obtaining a sequence of subimages automatically by processing a video recording of the subject. Extraction of the subimages forms part of the algorithm for tracking head movement which is briefly described later in "Animation by Subject Tracking."

The normal requirement for mouth image animation is to select sequences of subimages which produce an animation effect to follow some speech or to provide some facial expression such as a smile or frown. To record mouth expressions, a training sequence must be devised that contains all the necessary shapes. For speech animation the minimum number of mouth shapes required is surprisingly small. Height (1981), in producing a lip-reader trainer (also described by Myers, 1982), used a set of 19 visually distinguishable images, each containing a particular combination of lip, tongue, and teeth positions. He demonstrated that, from sequences of the images alone, it was possible for a reader to pick up the gist of a conversation. Table 8.1 shows how the 19 mouth shapes approximately relate to particular phonemic sounds. It should be remembered that the particular relationship between sound and shape depends to some extent on the accent and pronunciation of the speaker.

Subimage Storage Requirements

The animation process relies on the existence of a database containing subimages which are patched and blended into the main image in appropriate positions. The subimages which are to be stored are a series of eye images for both the left and right eye, and a series of mouth images.

The data storage requirements for individual subimages is easily determined. If it is assumed that the head and shoulders image of the subject fills the screen, measurements taken on the eye dimension of an average subject show that an eye subimage patch will occupy approximately 3 percent of the screen area. For the eyes, a single frame of video is thus able to store 33 subimages. The mouth subimage needs to be a little larger, needing perhaps 5 percent of the frame area and thus allowing 20 subimages per frame to be stored.

The storage requirements for realistic animation of the eye images are a function of the degree of animation required and the nature of that animation. A basic requirement is the ability to simulate an eye blink. For a fixed retina

Table 8.1. Relationships between mouth shapes and sound.

Round mouth shapes, minimal tooth exposure.	
b*u*t, boat, book, boot, *wh*ich, *w*it. *h*at, bet, boy. down, bob, b*u*y.	Least round Roundest

Round mouth shapes, teeth exposed.	
*y*ou, a*z*ure, *ch*urch, *j*udge, *sh*ut. b*i*t, ba*i*t, rent. bat, get.	Teeth together Teeth furthest apart

Wide mouth shapes, teeth exposed.	
*p*et, bot*t*om, *b*et, *m*et. *f*at, *v*at. *t*en, *d*ebt. bat*t*er, *s*at, *z*oo. b*ir*d. *k*it. *b*eat.	Closed (silence) Teeth furthest apart

Round mouth shapes, teeth and tongue exposed.	
*n*et, butto*n*. *l*et, batt*l*e. *th*at. *th*ing. about, roses nonspeech, laugh etc.	Least round Roundest.

position a realistic blink can be achieved using only four images. If the requirement is that the animated subject is to have the ability to swivel his or her eyes, a family of subimages must be provide to achieve this effect. In reality, because natural eye movements tend to be rapid, as few as five images will give a satisfactory result. For an animated subject to have an eye swivel ability combined with an ability to blink at any degree of eye swivel, a total of 20 images must be stored. For two eyes the total storage requirements amount to two video frames.

Subimage Blending

Mouth and eye animation is achieved by successively blending subimages into the base image. The blending operation is carried out before the texture mapping process described under the heading "Generating the Synthetic Image by Subject Mapping" is applied to produce the final image. Because the base and subimages are static frontal views, the areas of the base image into which the subimages are to be blended are known and fixed. In order to achieve smooth animation effects, it is important that each subimage is perfectly aligned within its bounding box. The size of the sub image bounding box is chosen so that, when the subimages are derived from a real subject, the facial areas outside the bounding box remain substantially unaltered regardless of the particular subimage being captured. In practice any sensible choice of bounding box size for mouth subimages always results in minor facial shape changes taking place outside the subimage bounding box area. Because of these minor facial shape changes in a real subect, direct pasting of the subimages onto the static facial picture always results in small visible discontinuities at the edges of the pasted areas. To eliminate these discontinuities, it is necessary to apply a blending function to produce a smooth transition between the base image and the subimage.

Figure 8.4 shows the bounding box of a subimage. A border width of B lines

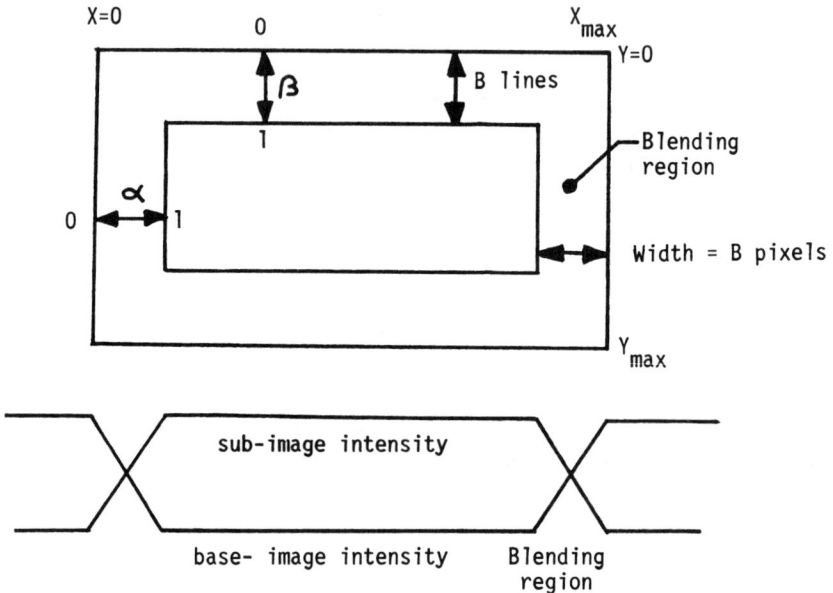

Figure 8.4. Subimage blending.

or pixels is defined around its perimeter. The procedure below carries out the blending function.

```
for Y = 0 to Ymax
  if Y < B then α = Y/B
  else if Y > (Ymax − B) then α = (Ymax − Y)/B
  else α = 1
    for X = 0 to Xmax
      if X < B then β = X/B
      else if X > (Xmax − B) then β = (Xmax − X)/B
      else β = 1
      γ = α * β
      pixel intensity at XY = (1 − γ) (base image
      intensity) + γ(subimage intensity)
    next X
next Y
```

The calculated value of the blending function γ has a value of 0 on the outer edge of the border and a value of one for the area of the subpicture which lies within the border. The effect of the blending function is illustrated in Figure 8.5. A mouth subimage pasted onto the main image without any blending is shown in Figure 8.5a; the border between the base image and the subimage is clearly visible. Where the blending function is applied over a border width of six pixels the boundary between the images disappears as shown in Figure 8.5b.

GENERATING THE SYNTHETIC IMAGE BY TEXTURE MAPPING

The base image used in the reconstruction process is a single frontal facial image. This image has to be mapped on to the 3-D facial surface defined by the vertices of the polygon model. The mapping function for this process must ensure that the image remains in registration with the underlying model regardless of the position and orientation of the final reconstruction. One method which is effective in performing this function is to maintain two polygon models. One of the models is held with its table of xyz vertex coordinates relating to a full frontal facial view. This model is held static, invariant to any transforms required to scale, orient, and position the final image. The second polygon model is initially upon derivation identical to the first. During the animation process this model undergoes the transforms necessary for the gross animation movement of the final reconstructed image. Figure 8.6 illustrates the process.

Hidden Surface Removal

The additional steps which are needed to map texture from the frontal image on to the transformed model during the animation process readily fit into the

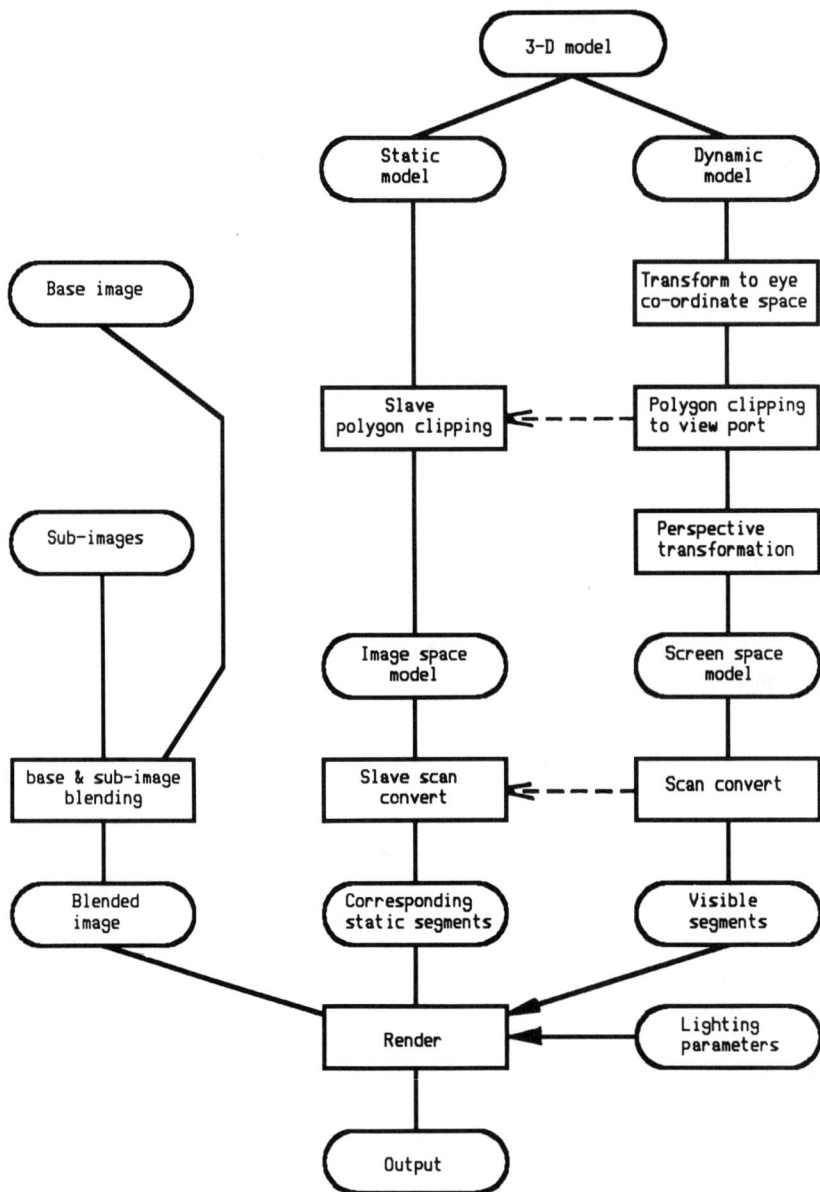

Figure 8.6. The texture mapping process.

algorithms which are required to carry out hidden surface removal on the underlying transformed model. The Watkins (1970) algorithm, which is used for the
illustrations in this chapter, is a scan line algorithm in which the final image is
written to the display one scan line at a time in a sequence going from the top

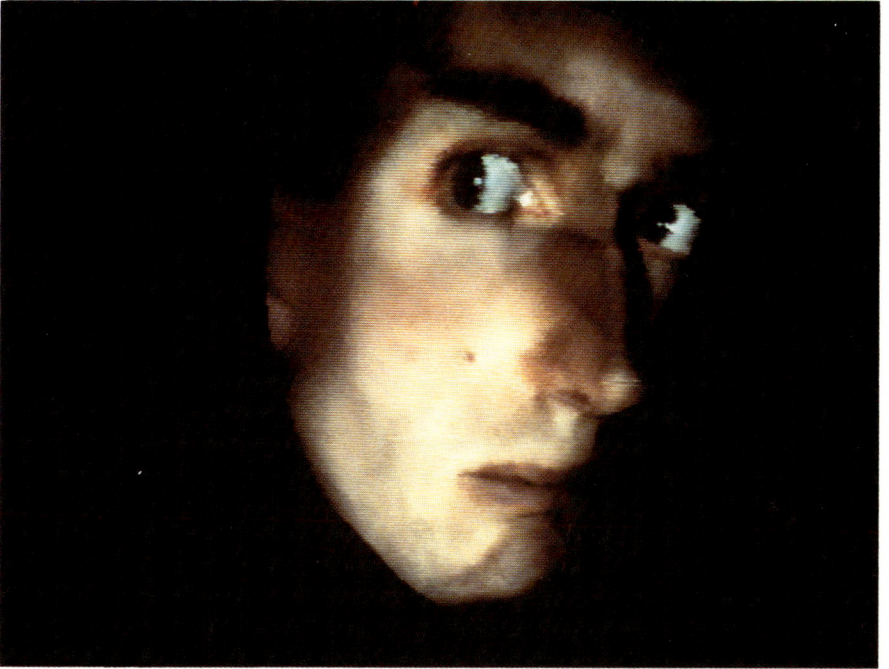

Figure 8.12(a) and (b). Charlie reconstructed with blended eye subpictures.

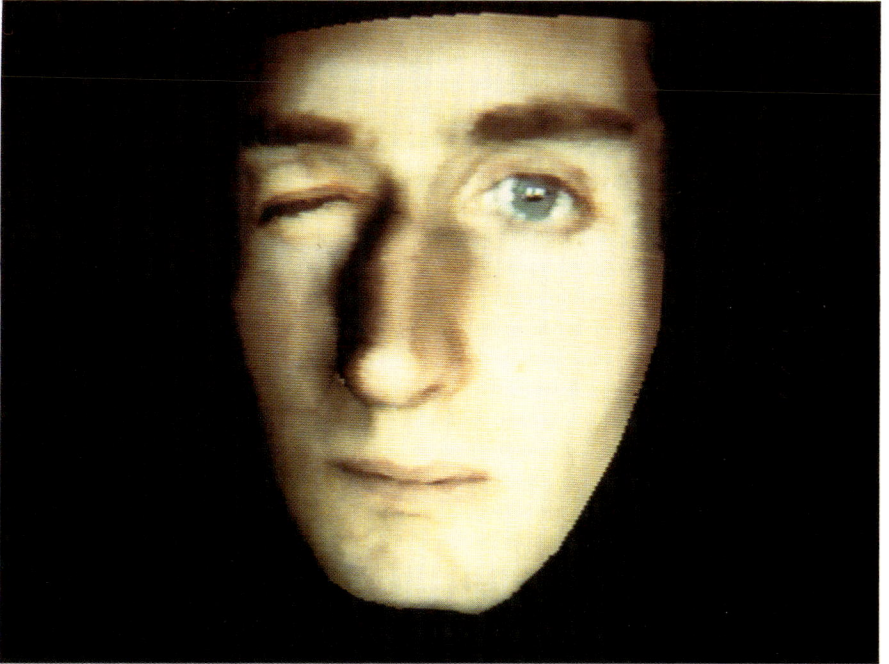

Figure 8.12(b).

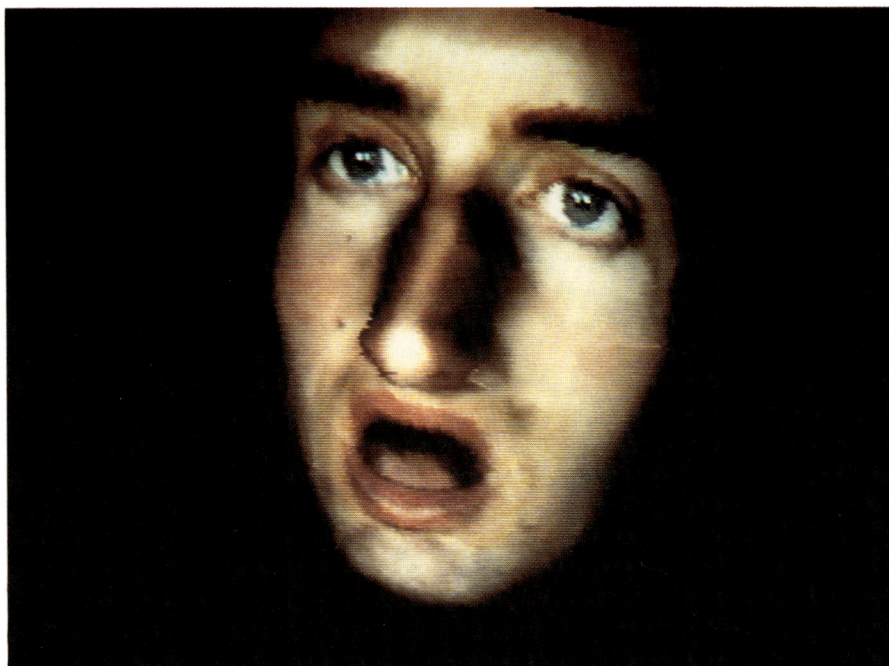

Figure 8.12(c). Charlie reconstructed with blended mouth subpicture.

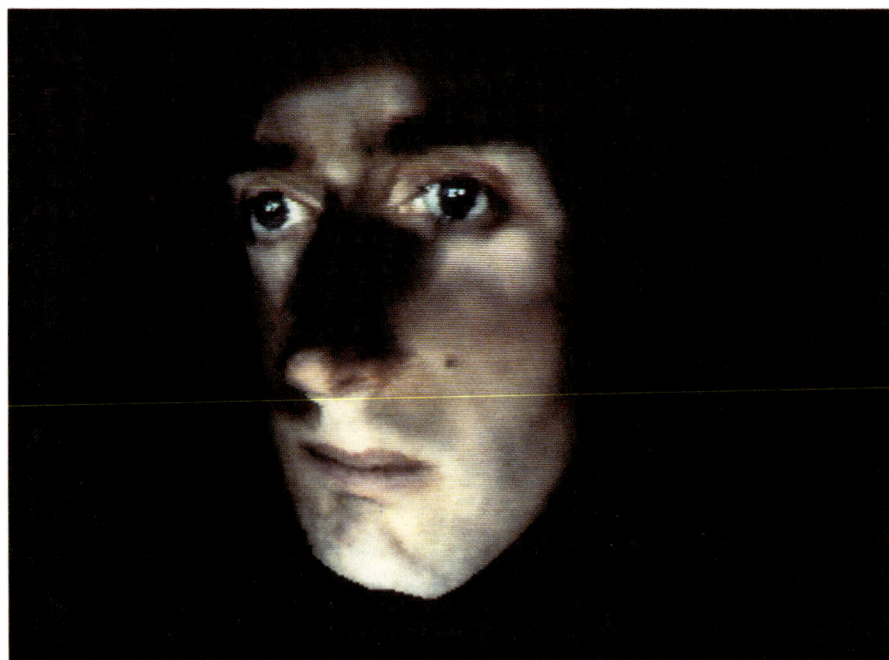

Figure 8.13. The effect of shadows from a simulated light source.

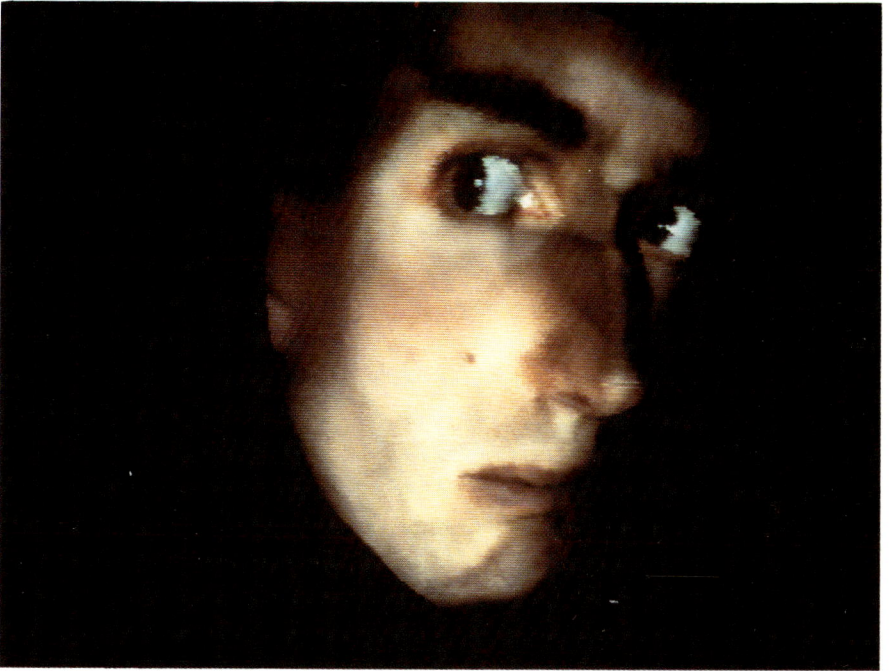

Figure 8.12(a) and (b). Charlie reconstructed with blended eye subpictures.

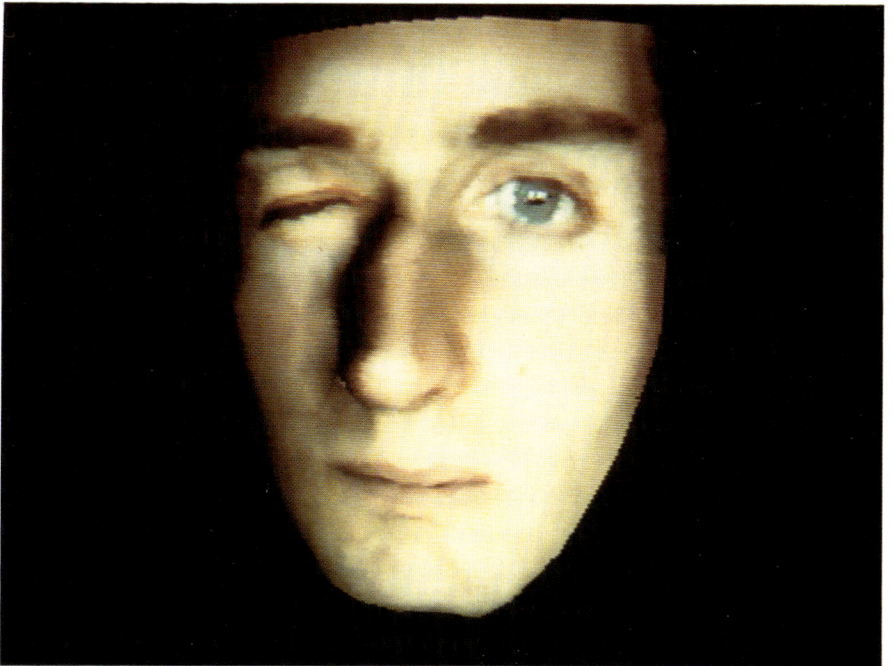

Figure 8.12(b).

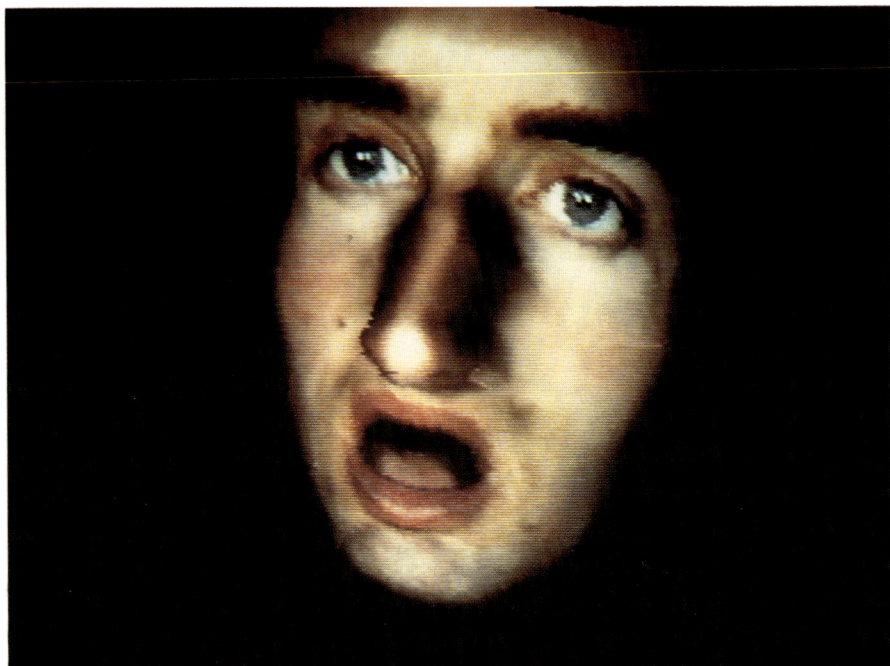

Figure 8.12(c). Charlie reconstructed with blended mouth subpicture.

Figure 8.13. The effect of shadows from a simulated light source.

or pixels is defined around its perimeter. The procedure below carries out the blending function.

```
for Y = 0 to Ymax
  if Y < B then α = Y/B
  else if Y > (Ymax − B) then α = (Ymax − Y)/B
  else α = 1
    for X = 0 to Xmax
      if X < B then β = X/B
      else if X > (Xmax − B) then β = (Xmax − X)/B
      else β = 1
      γ = α * β
      pixel intensity at XY = (1 − γ) (base image
      intensity) + γ(subimage intensity)
    next X
next Y
```

The calculated value of the blending function γ has a value of 0 on the outer edge of the border and a value of one for the area of the subpicture which lies within the border. The effect of the blending function is illustrated in Figure 8.5. A mouth subimage pasted onto the main image without any blending is shown in Figure 8.5a; the border between the base image and the subimage is clearly visible. Where the blending function is applied over a border width of six pixels the boundary between the images disappears as shown in Figure 8.5b.

GENERATING THE SYNTHETIC IMAGE BY TEXTURE MAPPING

The base image used in the reconstruction process is a single frontal facial image. This image has to be mapped on to the 3-D facial surface defined by the vertices of the polygon model. The mapping function for this process must ensure that the image remains in registration with the underlying model regardless of the position and orientation of the final reconstruction. One method which is effective in performing this function is to maintain two polygon models. One of the models is held with its table of xyz vertex coordinates relating to a full frontal facial view. This model is held static, invariant to any transforms required to scale, orient, and position the final image. The second polygon model is initially upon derivation identical to the first. During the animation process this model undergoes the transforms necessary for the gross animation movement of the final reconstructed image. Figure 8.6 illustrates the process.

Hidden Surface Removal

The additional steps which are needed to map texture from the frontal image on to the transformed model during the animation process readily fit into the

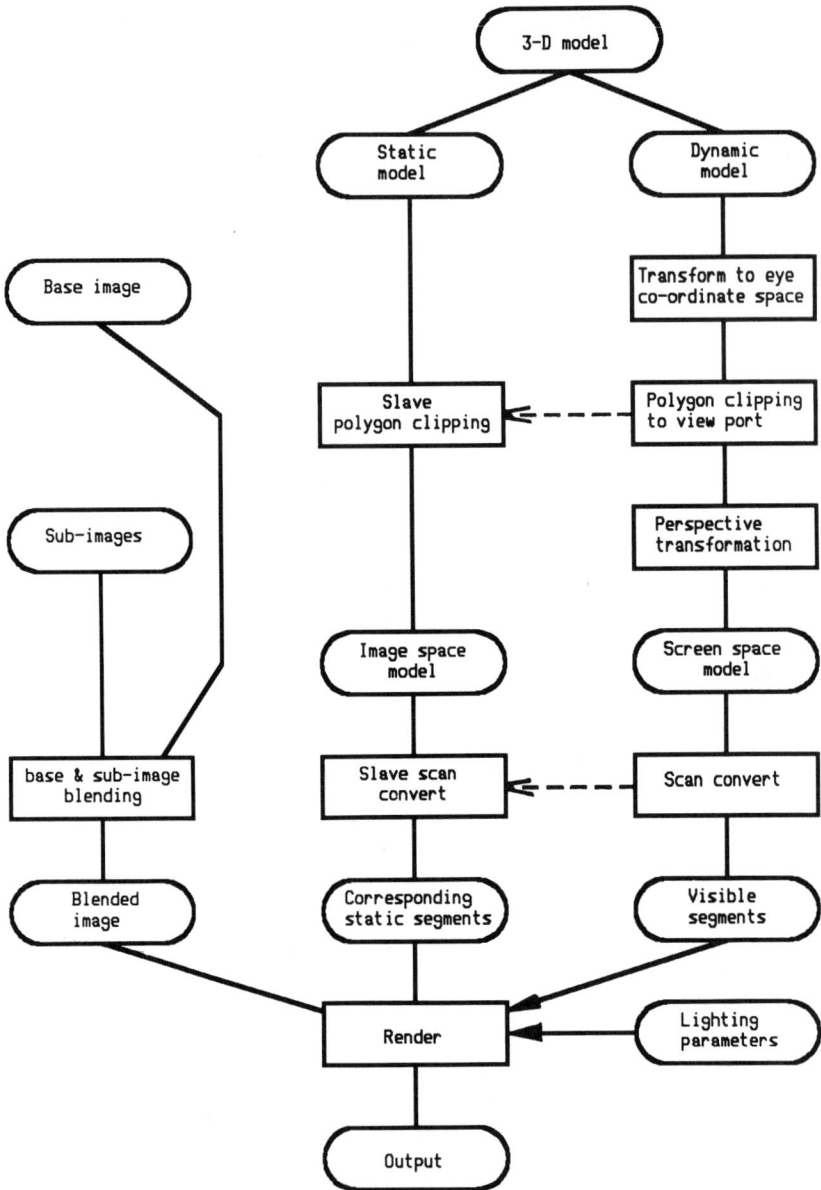

Figure 8.6. The texture mapping process.

algorithms which are required to carry out hidden surface removal on the underlying transformed model. The Watkins (1970) algorithm, which is used for the
illustrations in this chapter, is a scan line algorithm in which the final image is
written to the display one scan line at a time in a sequence going from the top

to the bottom of the display. Before the algorithm is applied, it is necessary to convert the image coordinate system from that used to define the 3-D model to the view screen coordinate system. This is carried out in two steps. First, a transformation is carried out to convert to the eye coordinate system. This transformation ensures that, where a rotation or translation is required for the next frame of an animation sequence, the direction of the z (depth) vector is in the eye-to-screen direction. A perspective transformation is then applied to convert from the eye to the screen coordinate system. The effect of this transformation is to alter the x,y, and z vertex coordinates of the model as a function of the z coordinate of the corresponding vertex in the eye coordinate system. The result of applying the transformation is that, when an object is viewed in the z direction into the screen, a perspective image is seen. These transformations are shown in Figure 8.7. Once they have been applied, the task of displaying the model is a

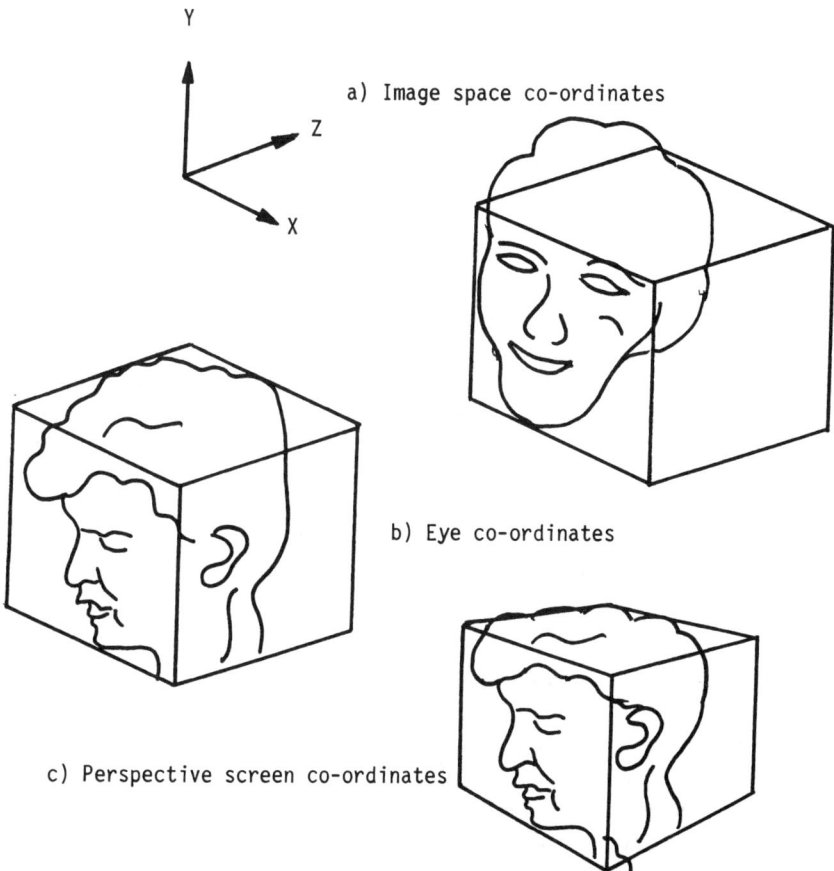

Figure 8.7. Coordinate systems.

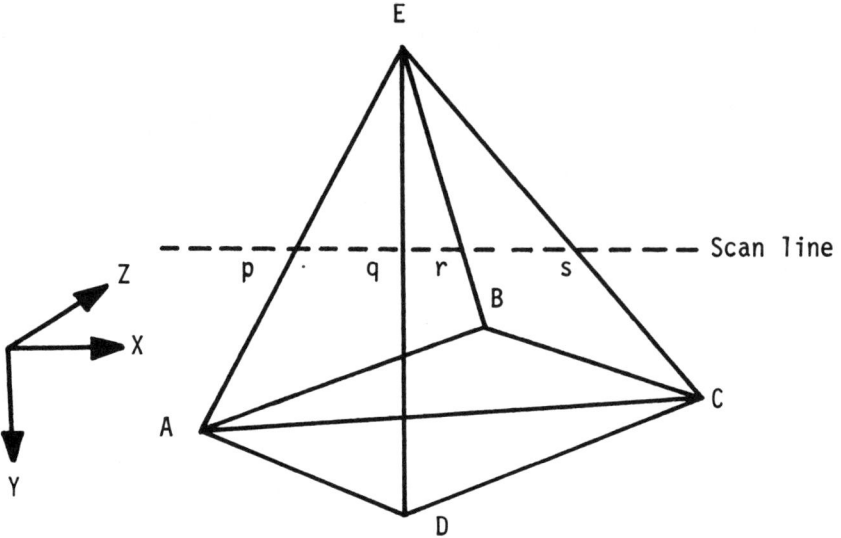

Figure 8.8. Hidden surface removal.

two-dimensional one in the xy direction. The z coordinate is used to determine which of the polygon surfaces are hidden behind other surfaces, and which will be seen by the viewer.

The standard Watkins hidden surface removal algorithm maintains a list of polygons which are crossed by the current scan line. For each line the points of intersection of the polygon edges and the xz plane of the scan line are calculated and stored in a list. Each polygon intersected by the scan line plane will have a segment created whose left and right endpoints have the same y values and whose x values are determined by the left and right intersection points. The segment of an intersected polygon is known as a *span*. Within each scan line there will be a number of spans, each created by the intersection of the scan line with a different polygon. In Figure 8.8 the spans are pr, pq, qs, and rs. In order to carry out hidden line removal, it is necessary to check the z coordinate (depth) associated with each span line, so that the spans which are near to the viewer and which are not obscured by other spans can be found.

Mapping from Dynamic to Static Model

Once a visible span on a scan line has been identified, the texture mapping process proceeds by identifying the corresponding span on the static model. Because this model is held in registration with the base image this strategy will ultimately allow pixels to be mapped from the base image. Figure 8.9 shows a visible polygon segment of the dynamic model in which the intersection of the

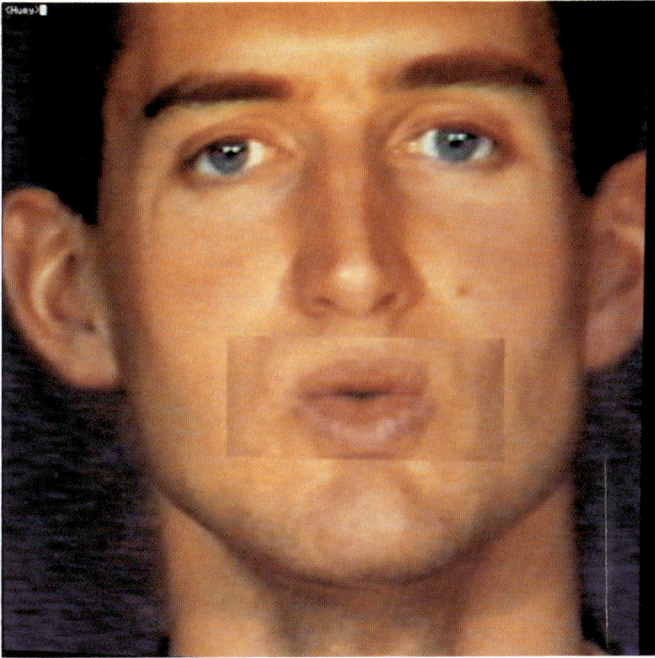

Figure 8.5(a). Pasted subimage with visible edges.

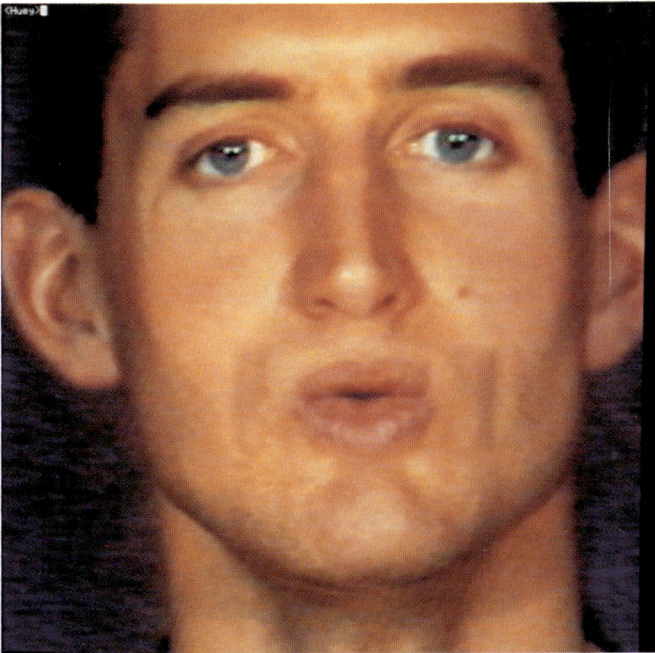

Figure 8.5(b). Application of the blending function smoothes the edges.

Figure 8.11(a). Original image of Zoe.

Figure 8.11(b). Zoe reconstructed.

scan line with polygon edges A_DB_D and A_DC_D are E_D and F_D, respectively. The distance A_DE_D along edge A_DB_D can be expressed as a proportion α of the edge length A_DB_D. By consideration of similar triangles, the constant of proportionality also applies to the distances in the x and y dimensions. Similarly, the distance A_DF_D can be expressed as $\beta\ A_DC_D$.

The xyz coordinates of the vertices A_D, B_D, and C_D of the polygon which is currently being scanned in the dynamic model have been obtained from particular positions in the dynamic model vertex list. The coordinates of A_S, B_S, and C_S, the corresponding polygons in the static model, are found in identical positions in the static model vertex list.

The dynamic model is held in perspective screen space where the xzy coordinates of the model have been modified by the perspective transformation according to the distance (z) of the original coordinates from the viewing position. A property of the perspective transformation is that lines and planes in screen space transform to lines and planes in the coordinate system of the static model. It is this property which makes it possible to interpolate between end points to find positions on the lines.

The static model is stored with its z dimension corresponding to depth. The intersections E_S and F_S on edges A_SB_S and A_SC_S of the static model correspond to the E_D, F_D intersection of the scan line and the polygon edges on the dynamic

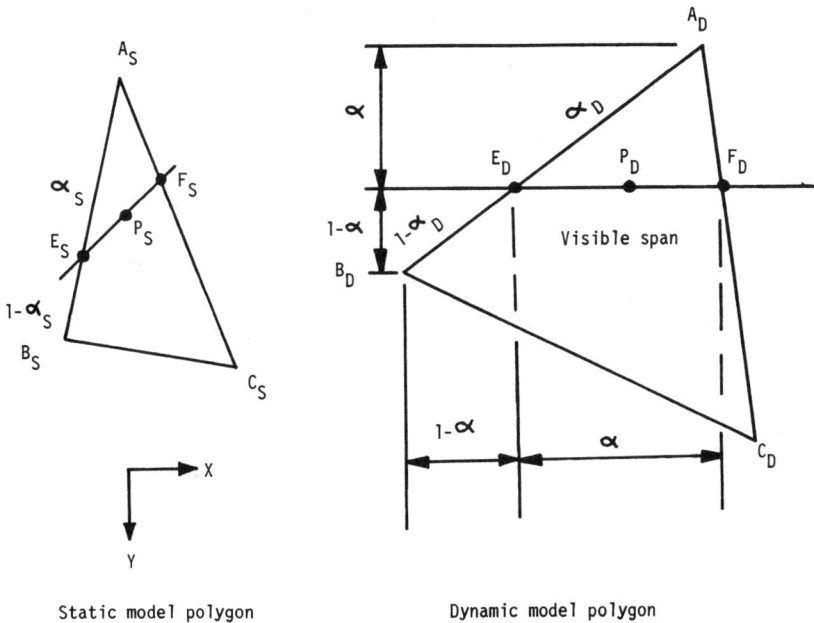

Static model polygon   Dynamic model polygon

Figure 8.9. Dynamic to static model mapping.

model. In order to enumerate E_S and F_S, the polygon edge interpolation factors α_D and β_D from the dynamic model are used to derive corresponding interpolation factors α_S and β_S for the static model. Newman and Sproull (1981) show that, when converting from perspective to normal coordinate systems, the parameters a_S and a_D are related by:

$$\alpha_s = \frac{\alpha_D * wa}{(1 - \alpha_D)wb + \alpha_D * wa}$$

$$w = \frac{Sz}{D}$$

Where
wa is the value of w at endpoint a
wb is the value of w at endpoint b
z is the depth of the appropriate endpoint in the eye co-ordinate system
S is the screen dimension
D is the distance between the eye position and the screen

The x and y coordinates of point E_S and Q_S on the static model are thus found.

Writing the Pixels

The line on the static model which corresponds to the visible segment on the current scan line has now been identified. Because the static model is held in registration with the underlying base image, a row of pixels along the line can be taken from the image beginning at point xE_S, yE_S and ending at point xF_S, yF_S. This row of pixels is written along the visible span on the scan line being processed. If the pixel position being written to the screen is P_D on the dynamic model, intensity information from corresponding pixel P_S on the base image must be written to the screen. If P_D is a proportional distance γ_D along visible span $E_D F_D$, a corresponding factor γ_S must be calculated using the formulae above. γ_S then determines the pixel position along the line on the static model. Thus, pixels are taken from the base picture and mapped on to the appropriate point on the image being reconstructed.

Pixel Undersampling and Oversampling

Because of the viewing transformations that have been applied to the static model to derive the dynamic model, the line on the static model and the corresponding visible span of the current scan line on the dynamic model will not normally be of

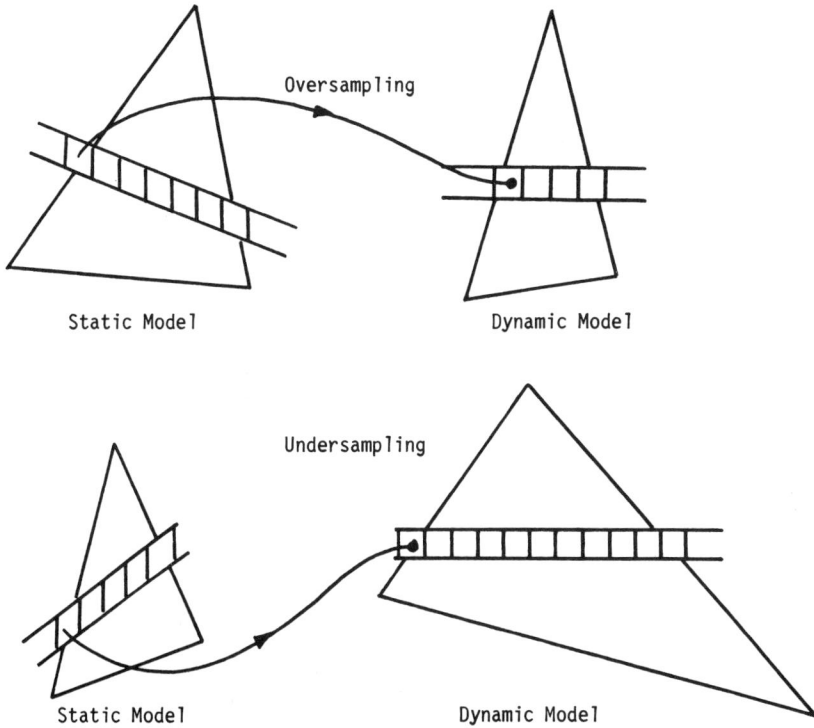

Figure 8.10. Pixel sampling.

the same length. As pixels are taken from the underlying image which is held in registration with the static model, an under- or oversampling will occur, as illustrated in Figure 8.10. Ideally, in order to avoid aliasing, the intensity of the mapped pixel should be derived from some averaging formulae applied to nearest neighbours of the calculated pixel position. In practice, because of the low spacial frequencies present in facial images, rounding the calculated x,y position to derive the integer coordinates of the pixel to be mapped does not appear to cause any significant undesirable visual effects.

Applying Shading

The appearance of the reconstructed image can be enhanced by combining the pixel intensity value obtained from the texture mapping process with that derived from a shading calculation. In any shading model, the intensity of light reflected from a surface point due to one or more light sources is calculated in terms of ambient, diffuse, and specular reflection coefficients. The two main shading methods in common use are those of Gouraud (1971) and Phong (1975), or derivatives of them. Both these methods fit efficiently into the hidden surface removal algorithm, and both depend on bilinear interpolation to compute reflection coefficients for points other than polygon vertices. Gouraud shading is computationally the less expensive of the two. Intensities are only calculated at polygon vertices. A bilinear interpolation scheme is used to provide values at each pixel. With Phong shading, surface normals are calculated at the vertices, and bilinear interpolation is used to derive normal values for each pixel position. The shading equation is then applied at each pixel to produce an intensity value. Although Gouraud shading suffers from disadvantages in its handling of specular reflections, this is of no great concern in the reconstruction of facial images, because the surfaces to be shaded are not shiny. The major difference between the techniques for this application is that Gouraud shading produces mach bands. These are visible bands of light on the illuminated model surface. The visibility of these bands is dependent on the coarseness of the polygon model, becoming more noticeable as the polygon size is increased.

CONTROL OF THE ANIMATION PROCESS

There are two separate processes required for animation using the texture mapping technique. The first is the manipulation of the facial polygon model; the second is the selection of the subimages which are to be mapped in sequence on to the base image. The manner in which both these processes are controlled depends on the application. For conventional story board animation, in which the animator determines the action of the reconstructed images, the parameters for specifying the images to be generated for each frame must be specified directly, or indirectly using key frame methods. In applications where the reconstructed image has to follow the motions of a live subject, a means of tracking the motion of the original subject must be used in which the tracking algorithm automatically generates the parameters required to animate the reconstructed image.

Conventional Animation

For story board animation the designer has to specify the image required for each frame in such a way that the frames fit together in a convincing sequence. In order to reduce the tedium of this task, key frame methods are often used whereby it is only necessary to specify the start frame and end frame of a sequence. Animation using the texture mapping process is readily amenable to key frame methods. For each frame of the image a set of parameters is required to define the image construction.

n	frame identifier
alpha	angle of field of view
vx,vy,vz	Viewing position
tx,ty,tz	Target position
amb	Proportion of ambient light.
lum	Simulated light source intensity
lx,ly,lz	Light source position
mx,my,mz	Head global movement
theta,psi,phi	Head orientation

For animation, sets of parameters are specified along the time axis to create the key frames. The number of frames to be created between the key frames is specified, and intermediate parameter sets are calculated by interpolating parameter values between those specified for the key frames. For a static base image the method will create a moving sequence of images each of which has a fixed facial expression. For animation of the facial features it is still necessary to specify the sequences of eye and mouth subimages. These images are blended into the base image before texture mapping takes place.

Animation by Subject Tracking

In an animation scheme in which the motion of a reconstructed image is to follow the motion of a live subect, the parameters for each video frame of the reconstruction must be derived from each frame of the original scene. The required parameters fall into two categories, the first to determine the particular subimages required to animate eye and mouth facial features, and the second to specify the position of the head image under six degrees of freedom. An algorithm to carry out these tasks has been described by Yau and Duffy (1989).

In the algorithm, feature locations are represented by boxes which fully enclose the features concerned. Assuming that the locations of the features have already been found in the first frame, the algorithm proceeds by using a block-searching process in which a sum of the squares of the differences of pixel intensity is used as a correlation measure between an image block in the current known position and image blocks in the frame being analyzed, which are at various displacements from the original position. The image block which has the best correlation measure is used as the new facial feature position. The tracking algorithm independently locates within each frame, the position of the bounding boxes of the nose, mouth, and eyes. As the head orientation is altered, the position of these boxes with respect to one another allows the parameters for head position and orientation to be derived. Note that positional parameter data does not require a high degree of accuracy. This is because, in any animated reconstruction, it is the dynamics of facial movement that convey the nuances of facial expression, rather than absolute position and orientation.

As so far described, the algorithm only tracks unchanging facial features. For a speaking subect this is obviously unsatisfactory. To overcome this difficulty the idea of having a code book containing a large number of prestored subimages is used. With this scheme, whenever a good correlation between a particular feature in a previously known position and its position in the frame being analyzed cannot be found, a search is made through the code book to find a prestored image which produces a better correlation figure. By this means the algorithm identifies the particular subimage which is the best match. The subimage identification code can be used directly in the reconstruction process.

One of the properties of the code book method is that it is able to generate the original code book entries rather than having to rely on their being predefined. If a search through the existing code book fails to identify any subimages which correlate well with the feature in the frame being analyzed, the feature is stored as a new code book entry. The algorithm thus provides an automatic method of generating the set of subimages required for animation.

APPLICATIONS

One of the main characteristics of the texture-mapping technique described in this chapter is that it allows the reconstruction of facial images by mapping real facial images onto an underlying 3-D model. This characteristic makes the technique ideal for applications where reconstructed images, animated or otherwise, are required to resemble real faces without any caricature- or cartoon-like overtones. Some application areas with these requirements are visual telecommunications, video speech synthesis, and forensic identification.

Visual Telecommunications

For teleconferencing or videophone applications, the transmission of video pictures requires a communication channel with sufficient bandwidth to carry the signal. Although much work has been carried out on coding schemes to reduce the required bandwidth, it does not at present seem likely that data rates of much less than 64K bits/second will be achieved. The 3KHz bandwidth of standard telephone circuits is clearly inadequate to handle such data rates.

For video phone applications, one way of reducing the bandwidth requirement is to reconstruct an animated image at the receiving end of a low bandwidth communications channel. A major requirement for this application is that the animated reconstruction is clearly recognizable as the speaker, and that its motion accurately follows the movements of the subject at the transmitting end. The reduction in channel bandwidth is achieved because the receiving equipment has a priori knowledge of the image to be reconstructed. This information comprises the data tables needed to define the 3-D polygon model of the subject's head, a single full facial picture and a collection of subimages which are to be texture mapped on to the complete facial picture. It is of course necessary to have some means by which this information is obtained in the first place, but even if the data have to be transmitted over the channel at the start of a conversation, they are only equivalent to a few video frames.

In the course of a conversation across such a channel, the information that would be transmitted would comprise the speech signal, positional information relating to orientation of the subject's head, and codes to define the current subpictures for the eyes and mouth. This information would be derived at the transmitting end by tracking the movement of the subject's head and by comparing the current eye and mouth subpictures to a set of subpictures which had previously been stored. A set of such data would be transmitted for each video frame.

At the receiving end, the data would be used to construct a frame-by-frame reproduction of the picture at the transmitting end. The positional information would be used to control the geometrical transformations required to set the face to the correct geometrical position in 3-D space. The codes for the subimages would be used to select the appropriate subimages for blending onto the underlying full face picture. The resultant full face picture would then be texture mapped onto the 3-D model to produce the final displayed frame. Thus, the final animated image would be produced on a frame-by-frame basis, with the reconstructed image in each frame being directly controlled by the action of a subject at the transmitting end of the channel.

Video Speech Synthesis

Modern telecommunications equipment is making increasing use of digitally prerecorded speech phrases and speech synthesis to give information to subscribers relating to call charges, alarm calls, call diversion information etc. With the advent of video phones, the texture-mapping technique will allow these messages to be accompanied by synthesized animated images whose movements are synchronized to the speech. In this application, the animation could be controlled either by a parameter table as shown in the section on "Control of the Animation Process," or possibly by direct analysis of the speech signal.

Forensic Identification

The storage of 3-D facial information, in addition to a full face photograph, makes possible the reconstruction of police mug shots from various viewing angles. This has obvious advantages where witnesses are attempting to identify a subject whom they may have seen from viewpoints other than that recorded on a conventional photograph.

A further application in this area is in the construction of 3-D photofit pictures. The blending of a base image and subimages, together with the use of a 3-D polygon model in which the vertex positions can be modified to produce various profiles, would result in a reconstruction that could be viewed from a number of angles. The ability to modify the reconstruction to reconcile views of the subject as seen by witnesses from different viewpoints may well lead to a more accurate description of the subject being obtained.

EXAMPLES

Some examples of the texture mapping process are shown in Figures 8.11, 8.12, and 8.13. These examples have all been processed using a Perkin Elmer 3230 running under Unix, and the results displayed on a Gresham Lion S214 frame store. Processing time for the images was 3 to 4 minutes for a texture mapped 256×256, 24-bit RGB image with Phong shading.

Figure 8.11a shows an original image of Zoe. Figure 8.11b is a reconstruction using only the base image together with a 3-D model obtained using the scanning technique described in the section on "Derivation of Surface Information." The blending of eye and mouth subimages is illustrated in Figures 8.12a,b, and c in the reconstructed pictures of Charlie. These pictures are single frames taken from a sequence animated using the key frame method described under the heading "Conventional Animation." The effect of augmenting the reconstruction by applying Phong shading from a simulated light source is illustrated in Figure 8.13.

REFERENCES

Aizaw, K., Yamada, Y., Harashima, & Saito, T. (1987). Modelling a person's face and synthesis of facial expressions. *Proceeding of the Globecom 87 conference IEEE Publication CH2520* (pp. 45–49).

Boissonnat, J.D. (1984). Geometric structures for three dimensional shape representation. *ACM Transactions on Graphics, 3(4)*, 266–286.

Duffy, N.D., & Yau, J.F.S. (1988). Facial image reconstruction and manipulation from measurements obtained using a structured lighting technique. *Pattern Recognition Letters, 7*, 239–243.

Gouraud, H. (1971). Computer display of curved surfaces. *IEEE Transactions, C–20(6)*, 623–629.

Height, R.L. (1981). Lip reader trainer: Computer program for the hearing impaired, *Proc. John Hopkins first national search for applications of personal computing to aid the handicapped* (pp. 4–5). Los Alamitos, CA: IEE Computer Society.

Myers, W. (1982, March), Graphics aid the deaf. *IEEE Computer Graphics and Applications*, pp. 100–102.

Nahas M, Huitric H., & Saintourens M. (1988). Animation of a B-spline figure, *The Visual Computer, 3(5)*, 272–276.

Newman, W.M., & Sproull, R.F. (1981). *Principles of interactive computer graphics* (2nd ed., pp. 361–362). New York: McGraw-Hill.

Parke, F.I. (1972). Computer generated animation of faces. *Proceedings of the ACM, 1*, 451–457.

Parke, F.I. (1982). Parameterised models for facial animation. *IEEE Computer Graphics and Applications, 2*, 61–68.

Phong, B.T. (1975). Illumination for computer generated pictures. *Communications of the ACM, 8(6)*, 311–317.

Waters, K. (1987). A muscle model for animating three-dimensional facial expressions. *Computer Graphics, 22*, 17–24.

Watkins, G.S. (1970). *A real-time visible surface algorithm* (Tech. rep. UTEC-CSc-70-101, NTIS AD-762 004). Salt Lake City, UT: Computer Science Department, University of Utah.

Yau, J.F.S., & Duffy, N. D. (1989). A feature tracking method for motion parameter estimation in a model based coding application. *Proceedings of the Third International Conference on Image Processing and Its Applications* (pp. 531–535). Warwick, England: IEE.

9
Modeling Three-Dimensional Facial Expressions*

Keith Waters
Schlumberger Laboratory for Computer Science
    Austin, TX

INTRODUCTION

The modeling of expressive three-dimensional faces is not a clearly defined task. Therefore, breaking down muscle activity into discrete modules is a valuable tool for the computer synthesis of facial expression (Waters 1987). *The Facial Action Coding System* (FACS), developed by P. Ekman and W. Friesen, psychologists of nonverbal communication, is a widely used notation for the coding of facial articulations (Ekman & Friesen, 1977). FACS describes 66 muscle actions (some by muscle blends) which in combination can give rise to thousands of possible facial expressions. These discrete units can be used as fundamental building blocks or reference units for the development of a parameterized facial muscle process. The face is capable of displaying a complex array of facial expressions; consequently, the control of facial muscles based upon FACS provides a valuable tool for the creation of grouped functional synergies which makes unnecessary the explicit control of individual muscles.

Isolating muscle actions from the three-dimensional facial topology avoids the hard-wiring of preformable actions. Therefore the development of a muscle process that is controllable by a limited number of parameters and is nonspecific to a facial topology allows a richer vocabulary and a more general approach for the modeling and animation of the primary facial expressions (Waters 1988).

Physically-based techniques for the modeling of nonrigid objects and their

---
* The initial research was supported by Middlesex Polytechnic Faculty of Art and Design. Thanks to Demetri Terzopoulos for explaining the principles of elasticity deformable models and for implementing the basic deformable lattice code.

motions are currently under investigation. The behavior of this class of deformable objects can mimic materials such as rubber, cloth, paper, and flexible metals and can simulate the melting of nonrigid solids into fluids by heat conduction (Terzopoulos & Fleischer, 1988a,b; Terzopoulos, Platt, Barr, & Fleischer, 1987). Human facial tissue, while physically complex, is also a deformable object which behaves in accordance with the laws of Newtowian physics. Consequently, it is desirable to apply some of the physical techniques to mimic the biomechanical properties of skin.

A combination of both a muscle process and a physically-based simulation of skin provides a valuable technique for creating synthetic facial expressions. However, it is the ability to manipulate the facial form, compute muscle activities, and display the image in real-time (rates higher than 10 frames a second) that provides the most powerful tool for facial imaging. To achieve a real-time display, a compromise must be reached between numerical stability and image fidelity. This load balancing between numerics and display, is not discussed in this chapter; however, future research will be implemented in this field (Pieper, 1989) that will provide a new dimension to modeling facial expressions.

FACIAL ANATOMY

This chapter considers the "face" as the frontal view of the head from the base of the chin to the hair line, and the frontal half of the head from the lateral view. These zones define the mobile areas of the head where facial expressions are formed. The following brief descriptions outline the physical characteristics relevant to computer synthesis.

Bone

The skull consists of 14 major bones, of which the mandible is the only freely jointed structure. Despite the skull's immobility, the facial skeleton comprises very discrete sections:

- The forehead consists of two bones called the frontal bones, which form the eyebrow ridge and the upper parts of the eyeball sockets.
- The eye socket, known as the orbit, consists of a number of bones: the ethmoid, lacrimal, and maxillary bones. They contribute to the generally conical space and the squarish outlets for the eyes.
- The zygomatic bone is a prominent feature that makes up the cheek bones.
- The maxillary structure forms the roof of the mouth and the floor of the nasal cavity. Its lower part supports the upper teeth.
- The mandible is a separate bone structure. It has a curved, horizontal body

which is convex forwards, and two broad rami which project upwards from the posterior ends of the body. The lower teeth are embedded in the upper surface of the mandible, and their importance should not be underestimated when modeling animated speech sequences.

The relative sizes, shapes, and distances between the mandible, maxilla, nasal, zygomatic, orbital, and frontal structures are almost infinitely variable, and it is this variation which uniquely characterizes each face. While the muscle and soft tissues change radically throughout life, it is the skull's structure that determines the general shape of the face we recognize.

Facial Muscle

The muscles that create facial expression are subcutaneous voluntary muscles. In general they arise from the bone and fascia of the skull and insert into a deep fascia layer which in turn is bonded to a layer of subcutaneous fat and skin. When the muscles are relaxed, the fatty tissues fill the hollows and smooth the transitions so as to allow the general shape of the skull to be seen.

The muscles of the face can be grouped according to the orientation of the individual muscle fibers and can be divided into the upper and lower face. Three types of muscle can be discerned as the primary motion muscles:

- **Linear/parallel**—muscles that pull in a linear direction, such as the zygomaticus major and corrugator.
- **Elliptical/circular**—sphincter type muscles that squeeze, such as the orbicularis oris.
- **Sheet**—muscles that behave as a series of linear muscles spread over an area, such as the frontalis.

With one exception, the muscles are attached to the skull at one end and are embedded into the soft tissue of the skin at the other. The exception is the orbicularis oris, one end of which is embedded in the group of muscles converging at the modiolus at the corner of the mouth.

There are two types of musculature contraction—isotonic, where the muscle shortens under tension; and isometric, where a considerable degree of tension may be produced but no shortening (Warwick & Williams, 1974). Facial muscle is isotonic in nature.

Skin

The skin is comprised of two layers—the dermis, covered by the epidermis. Beneath the dermis is the subcutaneous fat which is in turn bonded to the underlying deep fibrous fascia that is connected to muscle or cartilaginous surface.

The skin can be subject to considerable mechanical stress from internal and external forces. As a result the epidermis interface with the dermis is marked by ridge/grove interdigitations. This arrangement distributes forces over a large surface area and prevents the epidermis from being stripped off by shearing forces.

The skin behaves like a rubber sheet when articulated, deforming over and around the underlying structures. With the progression of age, the epidermis and dermis lose their elasticity, resulting in the deepening of flexure lines. These are either folds in the dermis associated with habitual joint movement, or lines of attachment to the underlying deep fascia.

Modeling the surface skin as a mesh affords each node a finite *degree of mobility* (DOM). The primary factors determining nodal mobility are:

- The tensile strength of the muscle and skin.
- The proximity of the skin to the muscle node of attachment.
- The depth of tissue at the node and its proximity to bone.
- The elastic bounds of the relaxed tissue, and the interaction of other muscles.

These characteristics can be utilized in the modeling of facial muscle and skin.

THE COMPUTER MODEL OF MUSCLES FOR THE FACE

The face is endowed with elastic, viscous, and other biomechanical properties that characterize the displacement activities of skin. The simulation of such interactions is not the objective of the Muscle Model Process. What is required is not the exact simulation of neurons, muscles, and joints, but a model with a few dynamic parameters that mimic the primary biomechanical characteristics. What follows is the description of the modeling of the three primary types of muscle: linear, sphincter, and sheet.

The parameters employed in the first muscle approximation (Waters,1987) were: the vector contraction, (a variable between 0 and 1), and the position of the head and tail of the vector describing the fall-off radius in three dimensions. These parameters did not provide the desired result, as linear muscles cannot yield a circular contraction towards a node. It was necessary therefore to establish, firstly, the area of flesh influenced by the muscle contraction; secondly, the length of the muscle; and thirdly, the position of the muscles in three-dimensional space relative to the underlying bone structures.

The zone of influence (the perturbed area of skin) depends upon the degree of muscle contraction. As the muscle tension increases, so does the area of skin influenced. Measuring real faces is at best difficult, as the range of surface characteristics varies greatly from face to face. The individual muscle contractions employed by FACS provide a source for an approximation of the zone of influence. Most muscles create a convex zone that varies from 15 to 160 degrees.

The three-dimensional computer muscle vectors were positioned in the facial model from anatomical descriptions (Ferner & Straubesand, 1983; Romanes, 1967). The computer muscle consists of an explicit position at the head and tail, which is located with a reasonable degree of accuracy, as the points of facial muscle attachment are usually spread over a small area.

Muscle Vectors

Muscles can be described with direction and magnitude, both in two and three dimensions. The direction is toward a point of attachment on the bone, and the magnitude of the displacement depends upon the muscle spring constant and the tension created by a muscle contraction.

In linear or parallel muscle the surrounding skin is contracted towards the static node of attachment on the bone until, at a finite distance away, the force dissipates to zero. In sphincter muscle, the skin tissue is squeezed towards a center, like the tightening of a string bag. This can be described as occurring uniformly about a point of contraction. Sheet muscle is a broad flat area of muscle fiber strands and does not emanate from a point source. As a result it does not contract to a localized node, but rather to a group of separated muscle fiber nodes. In fact, the muscle is a series of almost parallel fibers spread over an area.

The behavior of the linear, sphincter and sheet muscle results from low level muscle fiber contractions. Therefore the requirement is to specify the resultant displacement of an arbitrary surface node $\mathbf{p}_0$ to $\mathbf{p}_p$. This can be expressed by the following function:

$$\mathbf{p}_p \propto f(k, \beta, \ldots, \beta_n \mathbf{p}_o) \tag{1}$$

where k is a spring constant and $\beta, \ldots, \beta_n$ are variable parameters dependent upon muscle type. In this chapter, equation (1) is employed in the modeling of the three muscle types, and each muscle description requires complete specification of the variables $\beta, \ldots, \beta_n$.

Linear Muscle

For the linear muscle it is necessary to compute how the adjacent tissue, such as the node $\mathbf{p}$ in Figure 9.1, is affected by a muscle vector contraction. It is assumed that there is no displacement at the point of attachment (the bone) and that maximum deflection occurs at the point of insertion into the skin. Consequently, a dissipation of the force is passed through the adjoining tissue, both across the sector P_m, P_n and V_1, P_s.

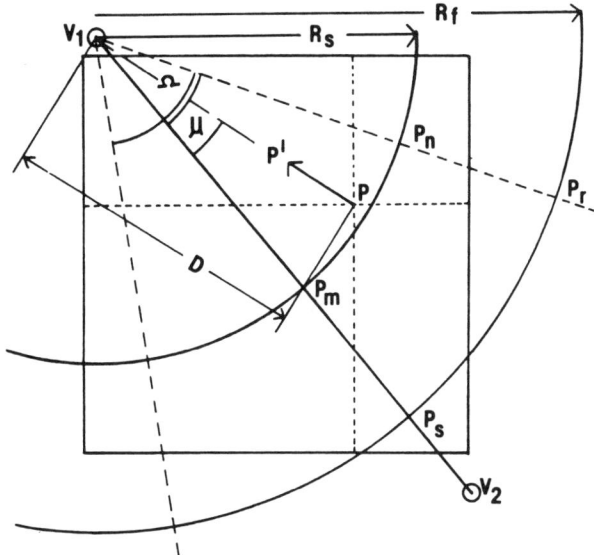

Figure 9.1. A linear muscle.

To calculate the displacement of an arbitrary node **p** located on the mesh to a new displacement **p'** within the segment $V_1 P_r P_s$, a displacement towards V_1 along the vector P, V_1 is created. Using equation (1):

$$\mathbf{p'} \, \alpha \, (k, a, r, \mathbf{p}). \tag{2}$$

Here the new location **p'** is a function of an angular displacement parameter

$$a = \cos(\mu) \tag{3}$$

where μ is the angle between the vectors V_1, V_2 and V_1, P and a radial displacement parameter.

$$r = \begin{cases} \cos\left(\dfrac{1-D}{R_s}\right); & \text{for p inside sector } V_1 P_n P_m V_1 \\[2ex] \cos\left(\dfrac{D-R_s}{R_f - R_s}\right); & \text{for p inside sector } P_n P_r P_s P_m \end{cases}$$

The Sphincter Muscle

The sphincter muscle can be described from a single point around which the surface is drawn together like the tightening of a string bag. This can be described as occurring uniformly about a point of contraction; consequently, the angular displacement is no longer required, and a major and minor axis are employed to describe the elliptical shape of the muscle (see Figure 9.2).

$$\mathbf{p}' \; \alpha \; (k, Lx, \; Ly, \mathbf{p}) \tag{4}$$

where k is the muscle spring constant, Lx represents the semimajor, and Ly the semiminor axes of the ellipse. The displacement of node $\mathbf{p} = [p_x, p_y]$ to $\mathbf{p}'$ in Figure 9.2 can be calculated using the function

$$f = 1 - \frac{\sqrt{Ly^2 p_x^2 + Lx^2 p_y^2}}{L_x L_y} \tag{5}$$

Sheet Muscle

Sheet muscle consists of strands of fibers which lie in flat bundles. An example of this type of muscle is the frontalis major, which lies on the forehead and is primarily involved with the raising of the eyebrows.

　　In terms of the basic function, an angular displacement is no longer required, as the muscle does not emanate from a point source and does not contract to a localized node (see Figure 9.3). In fact the muscle is a series of almost parallel fibers spread over an area. This can be defined as a displacement parallel to the direction of the central muscle vector. This can be defined as:

$$\mathbf{p}' \; \alpha \; (k, d, \mathbf{p}) \tag{6}$$

where k is the muscle spring constant and

$$d = \begin{cases} \cos\left(1 - \dfrac{L_t}{R_f}\right) ; & \text{for } \mathbf{p} \text{ inside sector } ABDC \\[2mm] \cos\left(1 - \dfrac{L_t}{R_f} * \left(\dfrac{V_i}{V_l} + V_f\right)\right) ; & \text{for } \mathbf{p} \text{ inside sector } CDFE \end{cases}$$

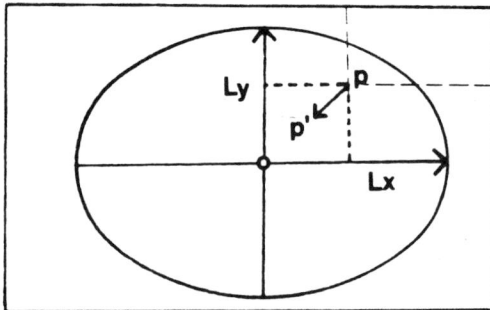

Figure 9.2. A sphincter muscle.

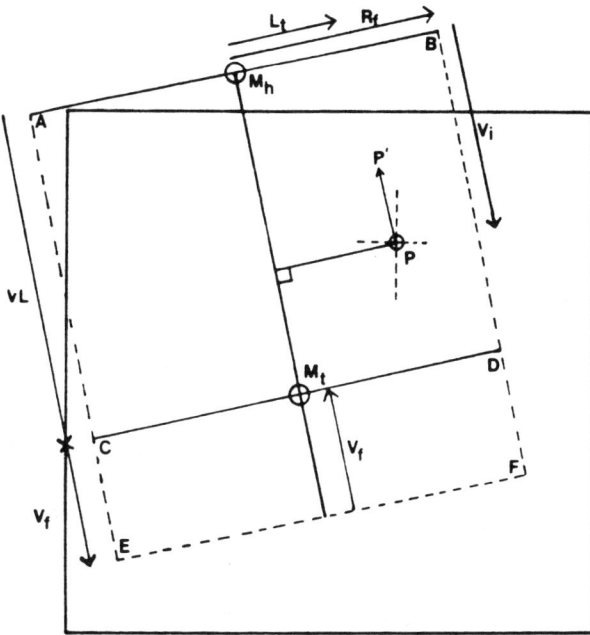

Figure 9.3. A sheet muscle.

Varying the Elastic Properties

In the three major muscle types described above the cosine function is used as a first order approximation to the elastic properties of skin. While this approach provides adequate results, it is evident that the elasticity of the face varies with age and from person to person. By replacing this function with a nonlinear interpolant, or Gaussian function, it is possible to vary the elasticity of the mesh (see Figure 9.4).

MODELING THE PRIMARY FACIAL EXPRESSIONS

Extensive research by psychologists of nonverbal communication has established a basic categorization of facial expressions generic to the human race (Goleman, 1981). Ekman, Friesen, and Ellsworth (1972) propose happiness, anger, fear, surprise, disgust/contempt, and sadness as the six primary affect categories. Other expressions such as interest, calm, bitterness, pride, irony, insecurity, and skepticism can be displayed on the face, but have not been as firmly established as fear, surprise, disgust, and sadness. The next section describes the construction of the six primary facial expressions (Ekman, 1971).

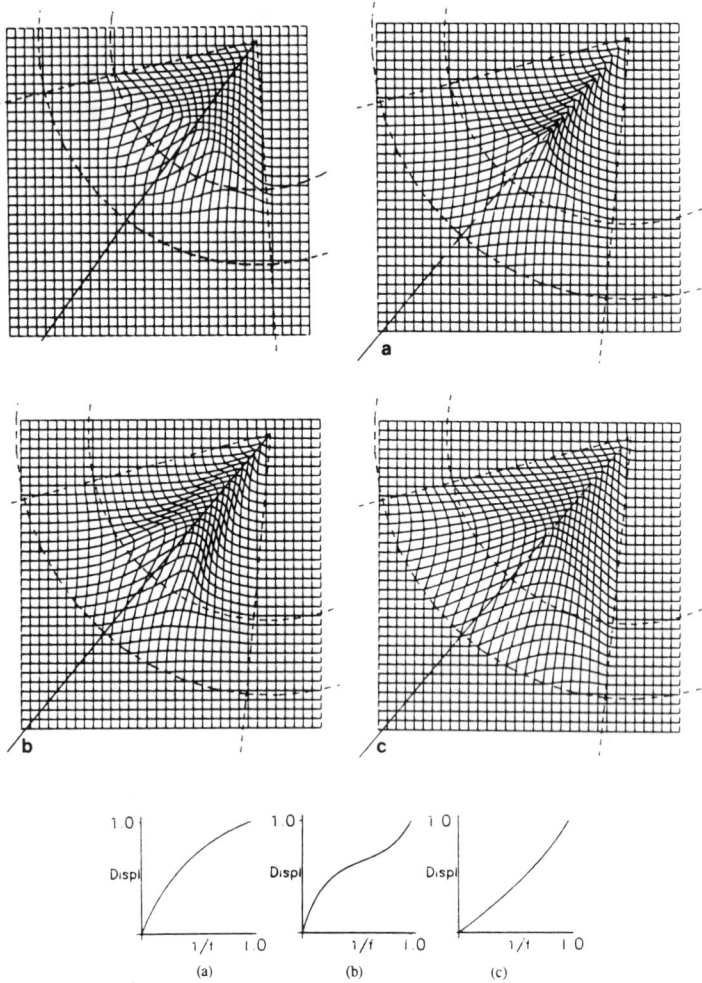

Figure 9.4. Varying the viscoelastic properties. Top left illustrates the muscle contraction cosine raised to a power 10.0. The remaining three illustrations are related to a variable Gaussian function variable between 0-1. Top right relates to the function (a), bottom left to function (b), and bottom right to function (c). All the contractions lay in the x,y plane with a zone of influence $\omega = 35.0$, fallstart $R_s = 7.0$, fallfin $R_f = 14.0$, muscle spring constant $K = 0.3$.

Surprise

Surprise is perhaps the briefest expression. In the upper face the brows are curved and raised (AU1 + AU2). There are no distinctive muscle actions in the

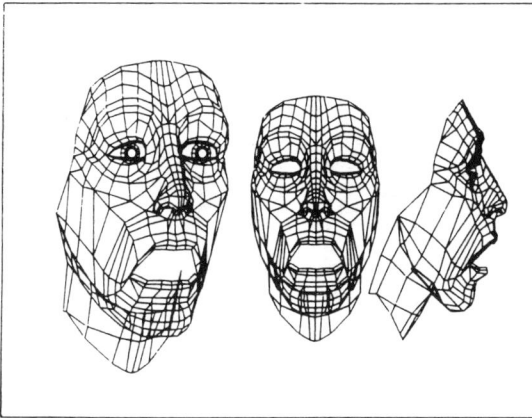

Figure 9.5. Surprise.

midsection of the face. The jaw drops, causing the lips and teeth to part. The more extreme the surprise, the wider the jaw becomes (Figure 9.5).

Fear

Fear varies in intensity from apprehension to terror. In the upper face, the brows appear raised and straightened (AU1 + AU2 + AU4). The eyes are tense during fear, with the upper lid raised and the lower lid tense. In the midsection of the face, the corner of the lips may be drawn backward (AU20), tightening the lips against the teeth. In the lower face the teeth are usually exposed by the downward pull of the lip (AU15 and/or AU16) (Figure 9.6).

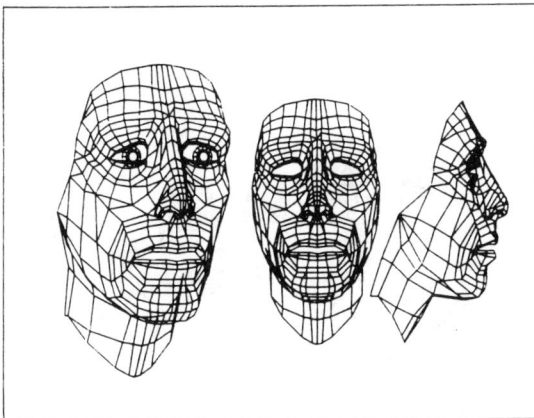

Figure 9.6. Fear.

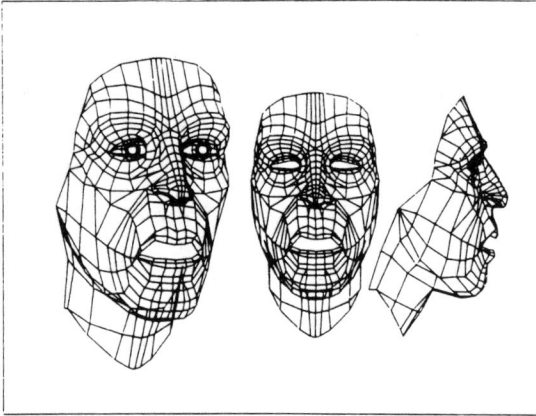

Figure 9.7. Disgust.

Disgust

Disgust is an expression of aversion, such as the taste of something you want to spit out. In the upper face there could be a lowering of the brows (AU4); however, the primary cues to the expression are found in the midregion of the face, around the nose and upper lip. Usually the upper lip is raised (AU9 and/or AU10), drawing up the flanges of the nose. The lower lip may be drawn downward or raised (AU17) (Figure 9.7).

Anger

In the emotional state of anger a person is most likely to harm someone purposefully. The brows are drawn down and together (AU2 + AU4), while the eyes

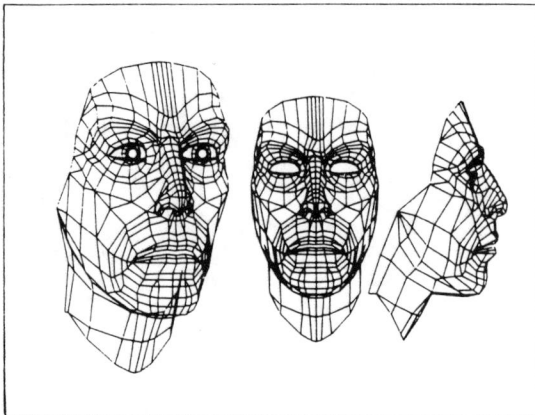

Figure 9.8. Anger.

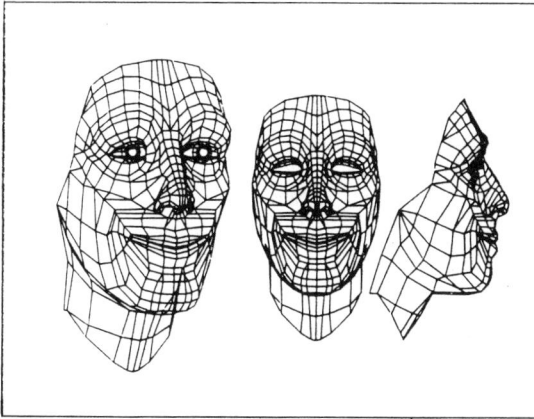

Figure 9.9. Happiness.

stare in a penetrating fashion with the eyelids wide (AU5). In the midregion of the face the flanges of the nose can be drawn upward (AU10). In the lower face region there can be two distinctive types of motion: the lips closed hard against the teeth (AU24), or the lips parted to bare teeth (AU25) (Figure 9.8).

Happiness

Happiness is a positive emotion, and can vary in intensity from mildly happy to joy or ecstasy. In the upper face the brows hardly change, while the eyelids are slightly compressed by the cheek raising up (AU6). The most prominent action is the raising of the corners of the lips that widens into a broad grin (AU12), and this is usually combined with deepening nasolabial folds (AU11) (Figure 9.9).

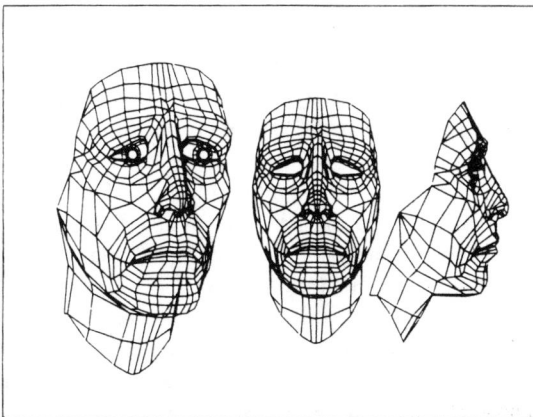

Figure 9.10. Sadness.

Sadness

Sadness is endured stress, and unlike surprise is often prolonged in duration. In sadness, the inner portion of the brows are drawn together and raised (AU1 + AU2 + AU4). The eyes are usually cast downward and the lower eyelids slightly raised. The mouth displays subtle motions that are akin to the expression of disgust where the corners of the mouth are pulled downward (AU15) (Figure 9.10).

EXTENSIONS TO SURFACE PATCHES

The techniques described are not limited to explicit polygonal representations. The face can be delineated as a collection of curvilinear patches, where each patch is specified by means of control points. Recent work by Waite (1989) employs bicubic B-splines for the modeling and animation of faces in a *Facial Action Control Editor*. For the modeling of the face, patches have two advantages over a direct polygonal description: firstly, the resolution of the model can be selected for the structure, and secondly, smooth blending of the surface articulations can be modeled. The muscle model process can also be mapped into this system to control the guiding polygon. The formulation therefore is once again an extension of equation (1):

$$\mathbf{p}'[k]i,j \propto f(k,a,r,\mathbf{p}[k]i,j) \tag{7}$$

where $\mathbf{p}[k]i,j$ is the set of control points describing the surface generated (see Figure 9.11).

Speech

Uttered speech consists of vowels and consonants. Consonants and vowels fall into two groups, called *visimes*. Early teachers of lip reading and visual speech perception, such as Jones (1918) at the turn of the century, further reduced speech into linguistic units known as *phonemes*.

Lip-reading is based upon the observation of some 45 English phonemes, not all of which are observable on the lips. Fromkin (1964) analyzed the mouth vowel shapes using frontal and lateral photographs, plaster cast, and X-rays. From the data collected, she concludes that, for the simplest models, one might use only two parameters to create vowel shapes in a physiological speech synthesizer—height and width of an ellipse, corrected so as to be equivalent to the actual area of the mouth opening of any vowel, and the protrusion being predicted from width. Further investigations by Brooke and Summerfield (1983)

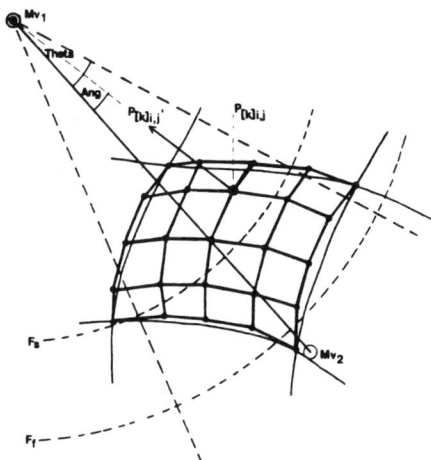

Figure 9.11. Bezier Patch Manipulation.

established that the width and area of the mouth opening to be paramount in the perception of a visible articulatory movement.

It is possible to determine the required muscle parameters for the muscle model process to distort an arbitrary mesh into any desired configuration in the same way that the mouth is distorted during speech. A recording on video tape was made of a person in fluent speech from the frontal and lateral views (Waters, 1988). Measurements of facial articulation were manually taken from this experiment, and a list of two-dimensional parameters constructed (see Table 9.1).

These parameters were also applied to a three-dimensional model, the principle difference being the protrusion and extension of the lips. The orbicularis oris is the muscle responsible for this action, through the constriction and pouting of the lips. Consequently, the squeezing action of the lips can be combined with the extension of the lips that is proportional to the width of the mouth. Figure 9.12 illustrates the same muscular contractions applied to a patch model. This correlation between the two- and three-dimensional models provides a valuable tool to speech synchronization, as the mouth shapes can be choreographed in two dimensions and then applied to three dimensions.

There appears to be an upper limit to the accuracy of matching mouth shapes to phonetics. As much as it would be useful to group actions together from phonetics, this can only provide a partial and impoverished solution to synchronization. The reasons for this are twofold: Firstly, phonetics are not created in fluent speech, and, moreover, phonetics do not take into account intonation and stress in speech. Secondly, there are perceptual problems because as the image fidelity of the face increases, we are more predisposed to compare the synthetic images with real people speaking and are unable to accept a visual and audio dichotomy.

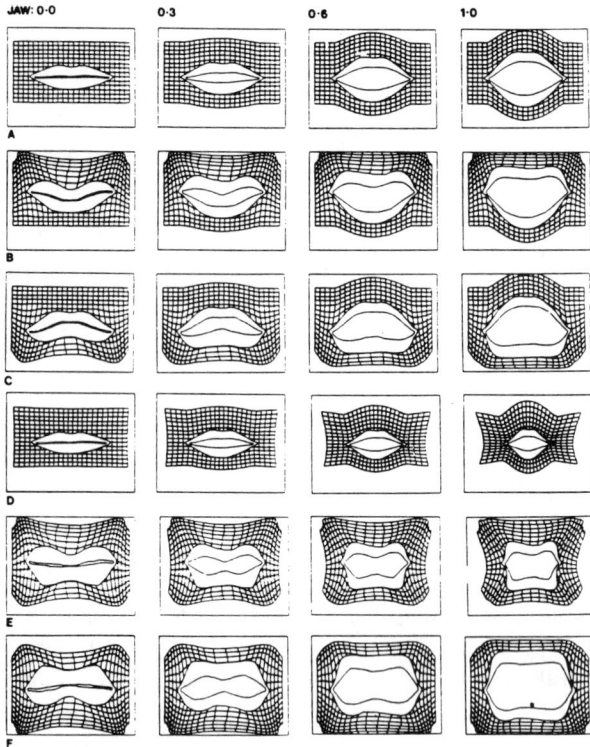

Figure 9.12(a). This figure illustrates the linear and sphincter muscles' actions in combination with the jaw opening on an arbitrary two-dimensional mesh. Row A illustrates (left to right) the jaw opening. Row B the jaw opening in combination with the zygomatic major contraction. Row C the jaw opening in combination with the anguli depressors. Row D the jaw opening in combination with the sphincter muscle. Row E the jaw opening in combination with the zygomatic major contraction. Row F the jaw in combination with the zygomatic major and the anguli depressors.

THE CONTROL OF FACIAL EXPRESSIONS

Biological motor behavior controls many animal actions, including human facial expression. Some attempts based on behavioral models have been made to control biological structures such as human locomotion and flock interaction (Reynolds, 1987), but for the most part there has been little evidence of computer animation based upon the underlying biological motivators.

Most human muscle fibers obey the all-or-none law which states that a stim-

Figure 9.12(b). This figure illustrates the linear and sphincter muscles' actions in combination with a Bezier patch mouth. Row A illustrates (left to right) the jaw opening. Row B the jaw opening in combination with the zygomatic major contraction. Row C the jaw opening in combination with the anguli depressors. Row D the jaw opening in combination with the sphincter muscle. Row E the jaw opening in combination with the zygomatic major contraction. Row F the jaw opening in combination with the zygomatic major and the anguli depressors.

ulus produces either a maximal response for that condition or no response at all. Further stimulation of a single motor nerve fiber of an intact muscle invokes a response in one motor unit, and all the muscle fibers of that motor unit respond (Landau, 1923). The motor unit therefore also obeys the all-or-none law. A similar principle of all-or-none has been exploited in a finite state approach to the synthesis of bioengineering control (Tomovic & McGhee, 1966). The system used a cybernetic actuator that has two binary inputs with one continuous output like a logical gate array to initiate the functioning of a mechanical limb. Gradations of contraction in a whole human muscle result from altering the number of active motor nerve fibers, and hence the number of motor units activated. A similar scenario can be used in the control of facial expressions.

Reflex Biological Control

The human biological motor system has evolved into an extremely complex structure capable of manoeuvering multilinked forms, with many degrees of freedom, through dynamic environments. The conscious decision of motion has been described as being built upon a base of reflex processes, or low-level motor programs regulated by high level controllers. The low-level motor programs, or Local Motor Programs (LMP), have been identified to be common principles for the control of walking in animals (Pearson, 1976) and may be the only economical method of controlling structures with multiple degrees of freedom (Zeltzer, 1984).

Motor programs can be observed in a variety of human activities, such as walking, grasping objects, and smiling (Warwick & Williams, 1974). The prehensile activity of the hand is a good example of motor activity. The power grip, the hook grip, and the precision grip have very distinctive characteristics where the group action of the fingers and thumb is orchestrated to perform particular acts. Likewise, facial muscles coalesce to form what we recognize as expressions.

Functional Synergies for Facial Expression

It has been demonstrated that the static expressions of happiness, anger, fear, surprise, disgust, and sadness can be generated from a limited set of muscles. The next task is to group these muscle subsystems to effect a particular class of time dependent facial expressions.

FACS provides us with a useful level of abstraction for facial animation. It deals with the low-level categorization of muscle actions. These activities can be grouped as in Figure 9.13(a) to create expressive facial animation by varying muscle parameters. However, this level the control of the expressions is tedious, as the animator is involved with the control of individual muscles.

Describing expressions as skills is a powerful abstraction because the low-level details of motion can be suppressed. By grouping the muscle actions as skills or tasks it is possible to specify expressions over time (see Figure 9.13(b)) and to be able to specify actions, such as, "be happy" or "be sad."

A DYNAMIC MUSCLE MODEL

So far this chapter has considered facial tissue as an infinitesimally thin membrane on which abstracted muscle actions can be performed (Waters, 1987; Magnanet-Thalmann, Primeau, & Thalmann 1988). This approach can produce unrealistic deformations, especially as multiple muscles contract in volumes of

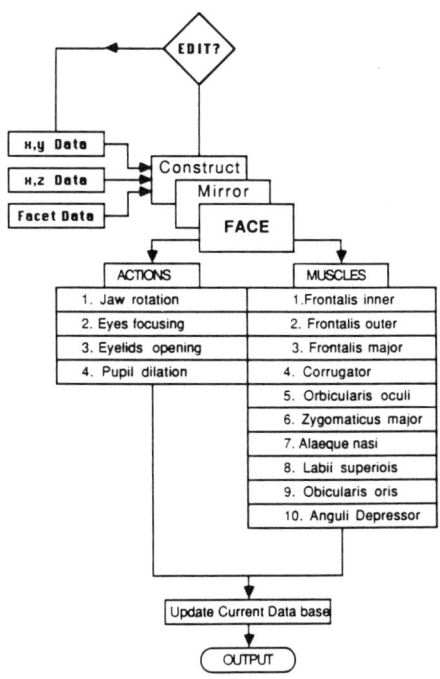

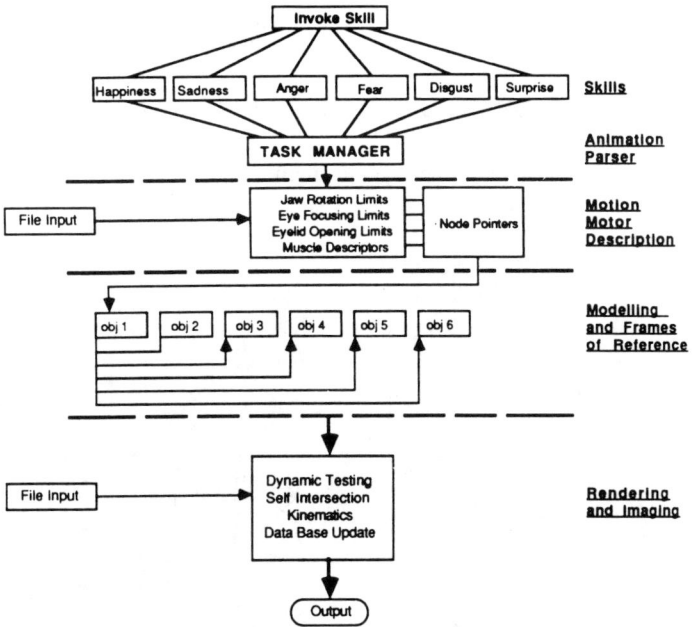

Figure 9.13. Parametric muscle control (a) and skills (b).

flesh. Therefore, it is considered merely a first order approximation to skin activity. A physical model of the epidermis, subcutaneous fatty tissues, and bone provides a more desirable and realistic technique for the modeling of facial expressions. The tissue construction can be represented by a dynamic constraint network with varying viscoelastic properties. Muscle force vectors are constructed from fibers that are rigidly attached to the bone at one end and blended into the dermis layer at the other. Finally, a numerical integration technique is employed to calculate the dynamic displacement of nodes in the lattice under the influence of internal muscular contractions.

Finite element techniques have been applied to approximate soft skin deformation in applications to plastic surgery and facial reconstruction. Larrabee reviews many of the techniques involved (Larrabee, 1986). Work by Deng into a finite element analysis of the surgery of human facial tissues also provides valuable techniques (Deng, 1988). Pieper (1989) employs a dynamic constraint spring network to model the mechanical behavior of skin that is similar to the techniques described in these notes.

Physically based techniques are computationally expensive, especially when considering more than a few thousand degrees of freedom (DOF). While facial expressions can be constructed from models containing only a few hundred DOF, it is desirable to create a higher level of image fidelity and more accurate biomechanical simulations (Moss, Linney, Gindrod, Arridge, & Clifton, 1987). Therefore, a two-level approach is suggested—firstly, a model to allow an interactive response and secondly, an overlay of dynamic simulation.

MEMBRANCE AND VOLUME LATTICES

A sheet (Figure 9.14) is a single layered membrane constructed from Hookean springs and node masses (particles) in u,v space. A discrete spring unit k linking

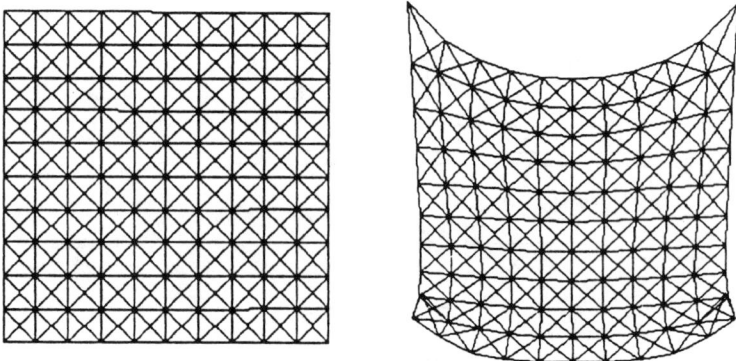

Figure 9.14. A spring mesh lattice at rest (a) and at equilibrium under the influence of gravity, supported at the four corners (b).

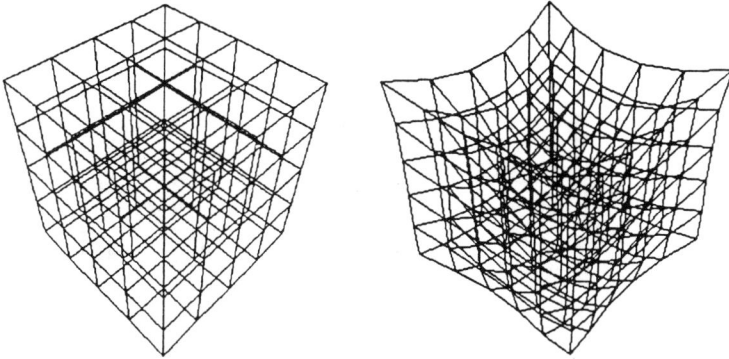

Figure 9.15. A static volume lattice (a) and the same lattice at an equilibrium poistion (b) under the influence of gravity being supported by the four corners.

node i at position x_i to node j at position x_j transmits both repulsive and attractive forces s_k to the linked nodes, such that

$$s_k = c_k(\| x_i - x_j \| - l_k) \tag{8}$$

where, c_k is the spring constant, l_k is the rest length of the spring, and $\| \cdot \|$ denotes Euclidean distance. To retain the sheet's structural integrity and resist shearing distortion in the u,v plane, the faces of the quadrilaterals are strutted, as shown in Figure 9.14(a) and Figure 9.14(b).

This membrane structure can be extended to a three-dimensional lattice network. Figure 9.15 illustrates a volumetric lattice in u,v,w space constructed from a base hexahedral element. While the basic hexahedral unit resists extension and compression in the u,v,w directions, it does not resist twisting and shearing forces. Consequently, this model is unstable and will collapse under the influence of external forces. By strutting each face of the hexahedron, the model resists extension, compression, torsion, and shearing in all directions and is therefore structurally stable.

Other stable units include tetrahedral and pentahedral pyramids. The advantage of these pyramids over hexahedra are twofold: firstly, they require fewer links to make them stable, and secondly, a combination of these units can be constructed to reflect the structure of skin tissue.

Figure 9.16 illustrates the construction of an extended lattice structure from base units to represent skin. The skin is visualized as an elastic volume constructed from layers of connected nodes. The top surface represents the epidermis, which is rather resistant to deformation due a stiff layer of kerantin and collagen. The second layer contains the muscle fiber attachments. The bottom layer of the model represents attachment to the nonmobile underlying structure of bone. The connections between the layers represent the subcutaneous fatty tissue

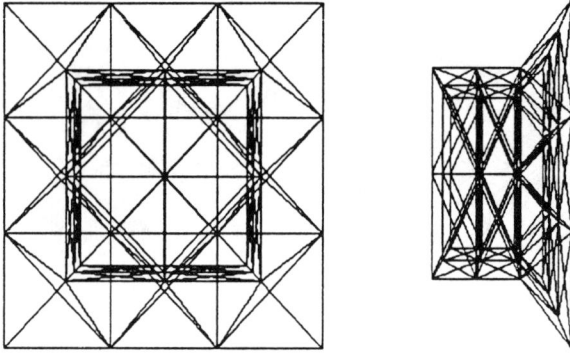

Figure 9.16. A Skin Lattice consisting of three layers.

and the ground substance, which is a semifluid medium. By modifying the spring constants c_k (1) of these interlayer connections, it is possible to mimic the behavior of elastic facial tissue.

Muscle Vectors

The muscle model process approximates the surface skin activity by calculating the displacement of nodes in a predetermined zone of influence. While this technique works well with single layered polygonal meshes, the behavior is unrealistic when volumes of facial tissue and bone are considered, because it is difficult to determine the displacement activities in a volume with varying structural properties. Therefore, a combined approach is employed to create a more biologically accurate model.

Muscles' fibers blend into the skin tissue at their mobile end, and when the muscle contracts the nodes are displaced towards the point of attachment to the bone. The muscle and skin coupling can be established from a zone of influence surrounding the muscle tail using the same technique as the muscle model process. However, a primary difference is that the muscles are attached to the layers between the bone and epidermis and not to the surface mesh. As a result, forces can propagate through the volume. The displacement of node j from $\mathbf{x}_j$ to $\mathbf{x}'_j$ due to muscle contraction can be computed as a weighted summation of all the muscle displacements, for $i = 1, \ldots .N$ acting on node j

$$\mathbf{x}'_j = \mathbf{x}_j + \sum_{i=1}^{N} c_i b_{ij} \mathbf{m}_i , \qquad (9)$$

where c_i are weights, and where $\mathbf{m}_i$ (10) is the muscle rest length, $\mathbf{r}_{ij}$ (11) is the computed distance to the tail of the muscle vector, and b_{ij} is the muscle blend

$$\mathbf{m}_i = \mathbf{m}_i^h - \mathbf{m}_i^t \tag{10}$$

$$\mathbf{r}_{ij} = \mathbf{m}_i^t - \mathbf{x}_j \tag{11}$$

$$b_{ij} = \begin{array}{ll} \cos\left(\dfrac{\|\mathbf{r}_{ij}\|}{b_l}\dfrac{\pi}{2}\right); \text{ for } \|\mathbf{r}_{ij}\| \le a_i & \tag{12} \\ 0; & \text{otherwise} \tag{13} \end{array}$$

where a_i is the radius of influence of the cosine blend profile.

Once all the muscle interactions have been computed, the lattice nodes $\mathbf{x}_j$ are displaced to their new positions $\mathbf{x}'_j$ (4). The result is that those nodes not influenced by the muscle contraction are in an unstable state, and forces propagate through the lattice until an equilibrium position is established. Figure 9.17 illustrates the stable rest state after the contraction of a muscle.

NUMERICAL TIME-INTEGRATION

To simulate the dynamics of the lattice model it is necessary to compute the node state variables, which include positions $\mathbf{x}_i$ and velocities $\mathbf{v}_i = d\mathbf{x}_i/dt$, in terms of forces $\mathbf{f}_i$ according to the second-order equations of Lagrangian dynamics (Terzopoulos et al., 1987):

$$m_i \frac{d^2\mathbf{x}_i}{dt^2} + \gamma \frac{d\mathbf{x}_i}{dt} + \mathbf{f}_i^e = 0 , \tag{14}$$

where m_i are masses and γ is the coefficient of velocity-dependent damping. The net forces at node $_i$ are

$$\mathbf{f}_i^e = \mathbf{f}_i^s + \mathbf{f}_i^g + \mathbf{f}_i^m, \tag{15}$$

where $\mathbf{f}_i^g$ are gravity forces, $\mathbf{f}_i^m$ are muscle forces, and

$$\mathbf{f}_i^s = \sum_{k \in C_i} \mathbf{s}_k \tag{16}$$

are spring forces (1) such that C_i denotes the set of springs connected to node i.

The Euler method may be used as a time-integration procedure, but it is only first-order accurate and can become unstable for modestly sized time steps. Stability may be improved by using a higher-order predictor corrector method such as the Adams-Bashforth-Moulton method (Press, Flanney, Tuekolsky, & Verttering, 1986); however, this can be computationally expensive. As a compromise solution to the numerical instability problem, a midpoint method, or second-order Runge-Kutta formulation, was employed. This compromise provided the required stability for the model.

The choice of damping coefficient is critical for the lattice to converge effectively. Fortunately, facial tissue, while being a flexible surface, does not oscillate about an equilibrium; rather, it behaves as if it were slightly overdamped, which improves the numerical instability.

THE CONSTRUCTION OF FACIAL GEOMETRIES

Volume lattices illustrated in Figures 9.17 are uniform solutions for the construction of a volume lattice. In reality the facial structure is doubly curved, with varying thickness and physical structure.

Despite these complexities, volume lattices can be constructed for the face starting with a single surface polygonal mesh model of faces. While perhaps not an elegant solution, this approach remains consistent with the volume lattice structures described earlier in this chapter.

The surface is first reduced to a triangular mesh. Normal vectors from the center of gravity of each triangle are projected downwards to establish nodes in the dermis layer. Tetrahedral units are then constructed from the triangle bases to these nodes. Finally, zero-length springs are attached to the mid layer and immobilized (in "bone") at the other end, so that the tissue structure can slide on the interface between the skin and the underlying bone.

Figure 9.18(a) is a static facial topology after lattice construction. A total of 960 polygons are used in the facial model, which results in approximately 6,500 spring units. The projected nodes are considered to be the dermis layer, into which the muscles blend and upon which the actions are performed. Figure

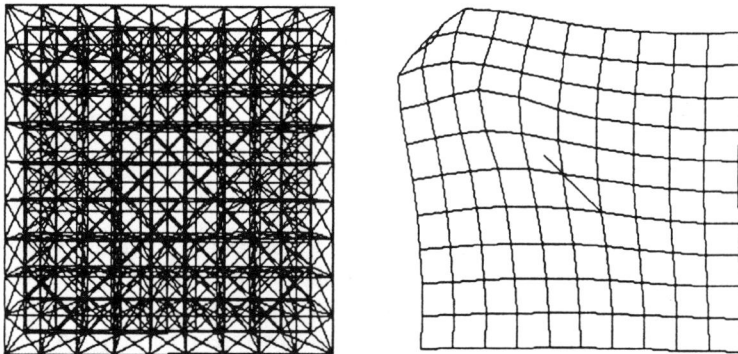

Figure 9.17. Skin Lattice before the muscle contraction (a) and after the muscle contraction with the base layers removed for clarity (b). Total number of springs = 1,802.

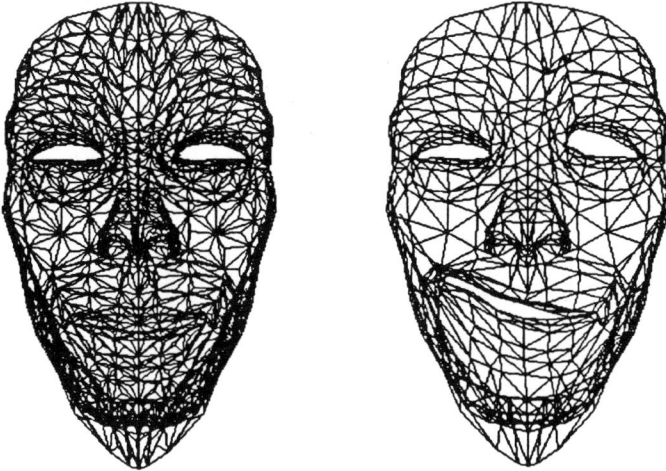

Figure 9.18. The lattice construction (a) and the deformation of the facial model under the influence of AU1 (left) and AU12 (right) (b) with the lower lattice links removed for clarity.

9.18(b) illustrates the skin distortion after the influence of the zygomatic major muscle (AU12), which raises the corner of the lip, and the inner frontalis muscle (AU1), which raises the inner portion of the brow.

CONCLUSIONS

The parameterized muscle model has been designed to perform complex facial articulations that correspond to a notational-based system. Moreover, it has been demonstrated to be a valuable tool for the creation of synthetic facial expressions based upon biological motivators rather than arbitrary selected parameters.

An approach for the control of facial articulations has been presented that overcomes the inherent complexities embedded in the degrees of freedom of the face. This has been designed to encapsulate the low level control with high-level commands.

Table 9.1.   Muscle parameters.

Action	Variable	Value Range
Jaw Rotation	Lip Separation	$0.0 \leq V \leq 5.0$
Oribularis Oris	Muscle Squeeze	$0.0 \leq V \leq 1.0$
Left Zygomatic	Muscle Tension	$0.0 \leq V \leq 1.0$
Right Zygomatic	Muscle Tension	$0.0 \leq V \leq 1.0$
Left Depressor	Muscle Tension	$0.0 \leq V \leq 1.0$
Right Depressor	Muscle Tension	$0.0 \leq V \leq 1.0$

The biomechanical properties of facial tissue have been modeled using physically-based techniques. As a result realistic static and dynamic images under the influence of internal and external forces can be produced. The combination of both the parameterized muscle process and a dynamic model has unified two important techniques that provides a starting point for the simulation of facial tissue in reconstructive surgery, the control of facial expression in psychology, and an accurate speech model to aid the deaf and hard of hearing.

REFERENCES

Brooke, M., & Summerfield, Q. (1982). Analysis, synthesis and perception of visible articulatory movements. *Journal of Phonetics, 11,* 63–76.

Deng, X. (1988). *A finite element analysis of surgery of the human facial tissues.* Unpublished doctoral dissertation, University of Columbia.

Ekman, P. (1971). *Unmasking the human face.* Englewood Cliffs, NJ: Prentice-Hall.

Ekman, P., & Friesen, W.V. (1977). *Manual for the Facial Action Coding System.* Palo Alto, CA: Consulting Psychologist Press.

Ekman, P., Friesen, W., & Ellsworth, P. (1972). *Emotion in the human face.* New York: Pergamon.

Ferner, H., & Staubesand, J. (1983). *Sobotta atlas of human anatomy. Vol. 1: Head, neck, upper extremities* (10th English ed.). Munich, Germany: Urban and Schwarzenberg.

Fromkin, V. (1964). Lip positions in American English vowels. *Language and speech, 7,* 215–255.

Goleman, D. (1981, February). The 7,000 faces of Dr. Ekman. *Psychology Today,* pp. 147–175.

Jones, D. (1918). *An outline of English phonetics.* Cambridge, London: Heffer.

Landau, B. (1923). *Essential human anatomy and physiology.* Glenview, IL: Scott Foresman.

Larrabee, W. (1986). A finite element model of skin deformation. 1. Biomechanics of skin and soft tissue: A review. *Laryngoscope, 96,* 399–419.

Magnenat-Thalmann, N., Primeau, E., & Thalmann, D. (1988). Abstract muscle action procedures for face animation. *The Visual Computer, 3,* 290–297.

Moss, P., Linney, A.D., Gindrod, S.R., Arridge, S., & Clifton, J.S. (1987). Three-dimensional visualization of the face and skull using computerized tomography and laser scanning techniques. *European Journal of Orthodontics, 9,* 247–253.

Pearson, K. (1976, December). The control of walking. *Scientific American,* pp. 72–86.

Pieper, S. (1989). *More than skin deep: Physical modeling of facial tissue.* Masters thesis, Massachusetts Institute of Technology.

Press, W., Flanney, B., Teukolsky, S., & Verttering, W. (1986). *Numerical recipes: The art of scientific computing.* Cambridge, England: Cambridge University Press.

Reynolds, C. (1987). Flocks, herds and schools: A distributed behavioral model. *Computer Graphics, 21(4),* 25–34.

Romanes, G.J. (1967). *Cunningham's manual of practical anatomy. Vol. 3: Head, neck and brain.* Oxford, England: Oxford Medical Publications.

Terzopoulos, D., & Fleischer, K. (1988a). Deformable models. *The Visual Computer,* *4(6),* 306–331.

Terzopoulos, D., & Fleischer, K. (1988b). Modeling inelastic deformation: Viscoelasticity, plasticity and fracture. *Computer Graphics, 22(4),* 269–278.

Terzopoulos, D., Platt, J., Barr, A., & Fleischer, K. (1987). Elastically deformable models. *Computer Graphics, 21(4),* 205–214.

Tomovic, R., & McGhee, R. (1966). A finite state approach to the synthesis of bioengineering control systems. *IEEE Transactions on Human Factors in Electronics, 7(2),* 65–69.

Waite, C. (1989). *The facial action control editor, face: A parametric facial expression editor for computer generated animation.* Masters thesis, Massachusetts Institute of Technology.

Warwick, R., & Williams, P.L. (1974). *Gray's anatomy* (35th ed.). London: Longman.

Waters, K. (1987). A muscle model for animating three-dimensional facial expression. *Computer Graphics, 22(4),* 17–24.

Waters, K. (1988). *The computer synthesis of expressive three-dimensional facial character animation.* Unpublished doctoral dissertation, Middlesex Polytechnic.

Zeltzer, D. (1984). *Representation and control of three-dimensional computer animated figures.* Unpublished doctoral dissertation, Ohio State University, Columbus, OH.

10

Faces as Surfaces

Vicki Bruce
Mike Burton
Tony Doyle
Department of Psychology
 University of Nottingham
 Nottingham, England

INTRODUCTION

The chapters in this second part of the book have each focussed on a different applied problem which involves the processing of facial images. Shepherd and Ellis discussed perhaps the most publicized of these problems, that of developing tools to help eyewitnesses reconstruct and recognize the faces of criminals. Linney described a system to allow plastic surgeons to plan their operations electronically. Duffy and Waters described progress towards the automatic animation of facial speech and expressions which could have an impact on the development of video telephones as well as on the world of commercial animation.

A common feature of the work described by Linney, Duffy, and Waters—in contrast to that described elsewhere in this book—is that the representation of the face used in these applications is of a three-dimensional surface rather than a two-dimensional pattern. In this chapter we will consider how understanding the face as a *mobile, bumpy surface* rather than as a *flat pattern* may have theoretical implications as well as practical applications, and towards the end of this chapter we will consider how such an understanding may in the future feed back to influence the development of forensic aids. In the course of this argument, we will also illustrate how our own research has benefited from current image-processing techniques. The experiments we mention depend upon good ways to manipulate and display representations of faces in both two and three dimensions.

Faces as Static Patterns

A common theme in studies of facial image processing (whether by psychologists or by computer scientists) is that the face is a flat pattern, like an alphanumeric character or geometric shape. This view has the consequences that individual variation between faces is considered only in terms of measurements made on the picture plane. We have seen examples of such an approach by computer scientists and engineers in Craw's chapter in the first part of this book.

An extreme example of this notion that measurements in the picture plane suffice to individuate faces was recently seen in a forensic context, in the trial of Michael Groce (son of Cherry Groce, accidentally shot when police raided her Brixton flat seeking her son). During the trial, which ended in March 1988, the prosecution compared two pictures: one of Michael Groce taken from a newsreel picture, and the other of a man captured on video as he robbed a building society. The prosecution claimed that these two pictures showed the same person. The pictures were individually quite indistinct, but an expert image-processing witness claimed that they showed the same man because, when the pictures were reduced to standard sizes, identical numbers of pixels separated key features on the two faces. The fact that the heads (both shown roughly in 3/4 view) were not properly standardized in terms of tilt or angle to the camera was apparently not considered. However, an expert witness for the defense (Alf Linney) pointed out the fallacy of the measures taken, and Mr. Groce was acquitted of the crime.

A second consequence of this view of faces as patterns has been in the design and development of electronic versions of reconstructive "kits" such as Identikit and Photofit. As we saw in Shepherd and Ellis's chapter, current electronic versions allow witnesses to manipulate the nature and placement of facial features very flexibly. However, the features are still construed essentially as two-dimensional parts embedded in a two-dimensional configural arrangement. To make a nose longer, it is moved down the face. Later we will consider a rather different approach which may have implications for the future design of such systems.

A final implication of the "faces as patterns" approach is that psychological experiments on the perception of faces, and attempts to automate face recognition on computer, have concentrated on full face views. The "caricature generator" (Brennan, 1985) described and extended by Benson, Perrett, and colleagues (see Benson, Perrett, & Davis, this volume) works only with true, full face views of faces. Research with the caricature generator has shown that distortions of the veridical form of a face can be recognized as well or even better than the face itself (Rhodes, Brennan, & Carey, 1987), consistent with the idea that faces may be represented in memory in terms of deviations from the norm or prototype face (Valentine & Bruce, 1986). Given that 3/4 views of familiar faces are recognized as rapidly as full face views (Bruce, Valentine, & Baddeley, 1987), then clearly the human visual system must somehow assess deviations from the norm more flexibly than the way implemented in caricature generators.

Another example comes from experiments on the relative salience of different face features, where subjects are asked to remember or compare faces in which one or other feature has been altered (see review by Shepherd, Davies, & Ellis, 1981, and recent papers by Haig, 1986a,b). In typical experiments, it has been shown that, of the internal features, eyes are more salient than mouths, which are in turn more salient than noses. However, two points should be noted. First, the shapes of noses are not very visible in a full face picture, compared with the shapes of eyes and mouths. Experiments using ³/₄ view faces might reveal a different pattern of salience, but as yet noone has investigated this, despite the fact that ³/₄ views are, if anything, better representations to use in experiments on perceiving and matching unfamiliar faces (e.g., Bruce et al., 1987). Second, despite the poor definition of the nose region from full face views, the nose region can be shown to be more important than the other internal features when the task is changed from matching identity to recognizing the sex of the face from photographs in which superficial cues such as hairstyle have been minimized (Roberts & Bruce, 1988). Which features are important from the face depends both upon the view shown and the task demands, and current research has not adequately considered the nature of the *problems* entailed by face processing, or the nature of the *stimuli*.

Faces as Mobile Surfaces

As has been well illustrated by several chapters in this volume, faces are not flat patterns but dynamic, bumpy surfaces that *grow* on underlying bony structures whose own growth is *constrained*. The orthodontist Enlow (1982) describes how these constraints have given rise to different facial types found between different races. The two extreme forms of head shape are the *dolichocephalic* head, which is long and narrow, and the *brachycephalic* head, which is wide and short. These different types of head in turn give rise to different face types—the *leptoprosopic* (long, narrow, and protrusive features) and the *euryprosopic* (broad, wide, less protrusive features). Because of the underlying head structures, facial features covary in these two extreme types of face, so that, for example, long, thin noses go with long, thin heads, and short, wide noses go with short, wide heads.

Understanding the true nature of the input to the face recognition system has its own consequences, both for theory and for practical applications. First, even if we wish to suggest that a perceiver does encode faces by measurements made at the level of the image (or "primal sketch," Marr, 1976; Watt, 1988; see also Watt, this volume), algorithms for the achievement of this will be developed more successfully given a proper definition of the input.

An excellent example of this is seen in the work of Pearson and Robinson

(1985). Pearson and Robinson attempted to produce an edge-finding algorithm that could accurately sketch a picture of a moving human face and hands, in order to compress gray-scale images of faces for economical image transmission in the development of videophones. Standard edge-finding algorithms produced unusable, cluttered representations, and this led them to devise a new edge-finding algorithm specifically tailored to the task of drawing sketches of faces. The algorithm that proved successful was one that looked for luminance *valleys* in the input, and its development was motivated by understanding that the places where lines need to be drawn in the image include locations where the *surface* of the face shifts sharply away from the line of sight.

If we suggest that the internal representation of a face comprises some as yet unspecified set of measurements from the primal sketch level, our understanding of what these measures should comprise may be helped by understanding the way in which different parts of a face may be mutually constraining during growth. For example, according to Enlow, the leptoprosopic face has a long, pointed head and a long, thin nose, while the euryprosopic face has a wide head with a short, wide nose. To the extent that overall head shape, nose shape, and perhaps some other features may *covary* in this way, there may be no need for the visual system to pay attention to the details of a nose when a "face-shape" measure taken from low spatial frequency information may covary with it. Alternatively, certain global parameters such as face shape may set bounds within which certain other properties are expected—and it may only be where some property violates this norm that it is noted (for example, the visual system might note "a long, thin nose for such a broad face"). Understanding the different constraints produced by the growth of face features, and gaining a formal understanding of the permissible or unlikely combinations of features, may have the additional advantage of allowing us to endow future electronic Photofit-like systems with some intelligent selection features, so that such systems could suggest likely features to accompany those that witnesses recognize confidently. The well-documented superiority of the police artist over either Identikit or Photofit (Laughery & Fowler, 1980; Davies, 1986; Ellis, Davies, & Shepherd, 1978) may arise because the artist has an implicit or explicit knowledge of such constraints, and future work in our laboratory is planned to examine this.

There is additionally, however, a more controversial implication of understanding faces as surfaces for understanding their recognition. It is possible that the viewer actually represents the surface shape of the face when constructing representations for recognition. Such a view is bolstered by evidence that line-drawings of faces (unless exaggerated through caricature—see Benson et al., this volume) are less well recognized than half-tone photographs which contain shading as well as "features" (Davies, Ellis, & Shepherd, 1978). Of course a line drawing itself conveys aspects of the surface shape—lines are drawn around the nose and around prominent facial features such as ex-President Nixon's "jowls." We are suggesting that we should think more carefully about what such lines tell us about the surface shape of the face, how such information might be enhanced by shading, and how such information may be represented in our internal descriptions of faces.

Further evidence for the use of surface-based information in recognition is reported by Bruce and Valentine (1988). They showed that, if the moving surfaces of faces of known individuals were displayed only by illuminating some 100 small spots of light on the surface (cf. Bassili 1978, 1979), then the faces were recognized at above-chance accuracy. If stills from the same videos were shown, no identification was possible, suggesting that the subjects recognized aspects of the shapes of the faces in 3-D as structures in motion (cf. Ullman, 1979). (We prefer this interpretation over one in which invariant information about identity is preserved in patterns of motion per se, since performance on this task was only slightly above chance and was no better from expressive motions—which might differ from one individual to the next—than from rigid motions of the face such as nodding and rocking the head.)

To summarize, understanding that a face is a dynamic surface, and understanding that it is the product of growth, may stimulate important theoretical and applied research directions. In this spirit, in a current project of research in Nottingham, we are investigating aspects of the three-dimensional structure of faces and heads, with the aim of specifying the nature of the global and local transformations to which a perceiver may be sensitive when categorizing faces. In the remainder of this chapter we will describe some of the first results we have obtained in research directed at the question of how subjects make relative and absolute judgments of the age and sex of a face.

JUDGING THE AGE AND SEX OF FACES

Age

Work within the framework of "ecological optics" (Gibson, 1979) has already been directed at specifying the nature of the information to which observers respond when judging the age level of heads. Pittenger and Shaw, and more recently Mark and Todd (e.g., Pittenger & Shaw, 1975a,b; Mark, Todd, & Shaw, 1981; Mark & Todd, 1985) have identified an abstract geometric transformation, cardioidal strain, which appears to characterize the change in the global shape of the craniofacial profile as it grows from infancy to adulthood. In a number of experiments using images like those shown in Figure 10.1, subjects have judged profiles towards the left of the series as younger in appearance than those towards the right.

While most of the work in this area has applied cardioidal strain to somewhat impoverished, two-dimensional representations of faces, Mark and Todd (1983)

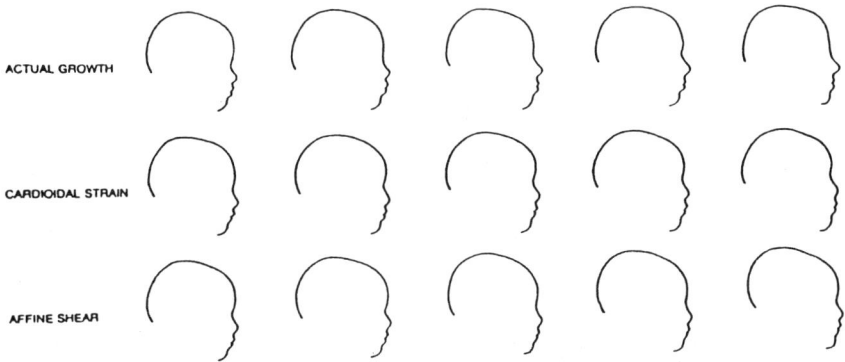

Figure 10.1. Examples of profile sequences resulting from actual growth, cardioidal strain and affine shear. From Mark, Todd, and Shaw (1981) and used with the first author's permission. Copyright 1981 by the American Psychological Association, Inc.

reported an interesting extension of this work in three dimensions. The algorithm for cardioidal strain expressed in spherical coordinates is:

$$R' = R(1 - k\cos\theta), \; \theta' = \theta, \; \phi' = \phi$$

where the free parameter, k, defines the extent of the transformation.

A computer was used to obtain an xyz coordinate representation from which to sculpt a bust of a girl aged 15 years. The database was then transformed, using the strain transformation, in a direction that should have made the girl appear younger in age, and a second bust was sculpted. The vast majority of observers saw the transformed bust as younger in appearance than the original, and when asked to assign absolute ages to the two versions, assigned an age of about 14 years to the original, and about 6 years to the younger head. While Mark and Todd's (1983) demonstration clearly showed that the strain transformation could be extended to three dimensions, it left open the question of whether it was really perceived in three dimensions, since subjects were free to view the pair of heads in any way they chose, and might have based their judgements on the shapes of coincident occluding contours. Additionally, it should be noted that the transformation has the effect of altering the slant of the lower half of the face: as the transformation is applied in the 'younger' direction, the chin is pulled upwards and inwards. The busts in Mark and Todd's experiment did not have the slant of their faces standardized and this might have contributed to the impressions of relative age which resulted.

In our own research (Bruce, Burton, Doyle, & Dench, 1989) we have made use of a graphics workstation to display three-dimensional models of heads obtained by laser scanning (by Alf Linney's group, in the manner described in his chapter). We wished to explore more carefully whether subjects could extract information about relative age level in three dimensions, by presenting subjects with pairs of faces where viewpoint was controlled and where the slant of the face could not be used to make the judgments. From a basic database whose origin lay in the middle of the head (on a line passing through the earholes), cardioidal strain was applied in a direction which should make the resulting head appear younger. The free parameter, k, was set to -0.07, $-.14$, $-.21$, and $-.28$, so that four further versions of the head were produced (see Figure 10.2). Subjects were shown pairs of heads which could differ by one "step" in age (e.g., the two versions produced when k $= -.14$ and $-.21$) through to four steps (the two versions k $= 0$ and k $= 0.28$). The two views shown could both appear in the same view (e.g., both profile or both ¾ view) or they could show different views (one profile and one ¾) (see Figure 10.2). Where the two views shown were the same, subjects could compare the shapes of the occluding contours directly in order to make their judgments, while, when the two views differed, no such strategy would be available. We were interested to learn whether subjects were more sensitive to relative age level when shown pairs of the same view compared with the mixed view condition.

We found that subjects were as good at making the relative age judgments to mixed views as they were to same view pairs (see Table 10.1), suggesting that subjects were indeed able to extract age level in three dimensions. However, we also found that subjects were rather inconsistent in their relative age ratings unless they were given "constrained" instructions which directed them to consider the differences in shape between adults' and infants' heads and to choose the younger or ("more babyish") in shape. In our studies we were experimenting with transformations to a clearly adult head with some rather strange features such as apparent baldness (though all subjects were told that the model wore a bathing cap), and it may be that such materials gave rise to a broader set of strategies than is available for the more impoverished materials used in much of

Table 10.1. Mean percentage correct responses made by subjects directed to choose the younger ("more babyish") head from pairs shown at different levels of cardioidal strain, as a function of the number of "steps" of strain level separating them, and the viewpoints shown.

STEPS:	1	2	3	4	MEAN
VIEWS:					
Profiles	68	74	81	83	76
3/4 views	77	82	87	88	84
Mixed	65	77	83	85	78

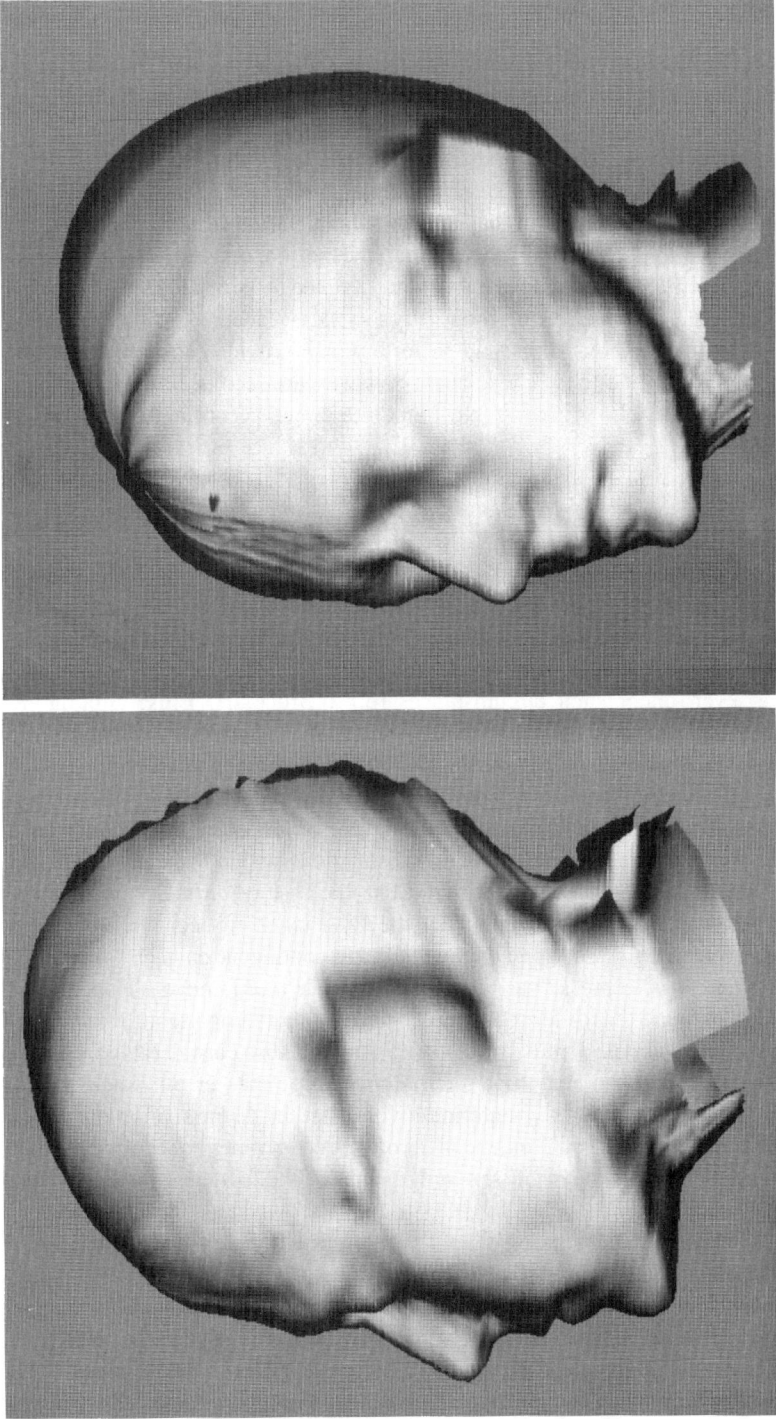

Figure 10.2. The effect of cardioidal strain on a 3D graphical image of a head. The left panel shows a profile view of the untransformed head, and the right shows a ¾ view of a head age-transformed in the "younger" direction. with k = −0.21.

this work. This sensitivity to instructional set has also been noted by Mark et al. (1981) in work where subjects were shown a set of profiles like those of Figure 10.1 and asked to describe the style of change that characterized the transformation from left to right. Subjects were much more likely to mention "growth" if this had been given as an example of the styles that might be present. Thus our experiments support the notion that cardioidal strain level may be perceived "in 3-D" while at the same time questioning the notion that this is an abstract invariant to which perceivers are "directly attuned." We prefer to think that cardioidal strain has an effect on the relative vertical placement of face features, shape of the forehead and chin, and so forth which provide cues which may be used in the judgment of age. These cues, however, are seen in combination with others (hair style, skin wrinkling, etc.) and interpreted accordingly (cf. Mark et al., 1980).

Sex

Deciding what sex a person's face is seems effortless, yet it is difficult to pinpoint how we make such judgments when superficial cues from hairstyle, make-up, or beards cannot be used. Yet subjects are remarkably quick and accurate, even under such circumstances (e.g., Bruce, H. Ellis, Gibling, & Young, 1987; Roberts & Bruce, 1988). H. Ellis (1986) suggested that identifying a person's sex was an early stage in the identification of the face which proceeded through a hierarchy of progressively more refined perceptual tests. The process would first lead to a decision that this was a face, then decide its race, sex, and age, and finally make finer individuating distinctions within these broad categories. However, Bruce et al. (1987) found that the ease or difficulty of deciding that a face was male had no effect on the time taken to decide that he was familiar, even when these judgments were made contingent on each other (i.e., when the task was to press "yes" if the face is male *and* familiar). Roberts and Bruce (1988) found that different features appeared more important for the determination of sex than of identity, since concealing the nose considerably impaired sex judgements but did not impair familiarity judgments at all. Such results therefore suggest, somewhat counterintuitively, that the determination of the sex of a face proceeds via a route independent of that concerning its identity. Further understanding of the basis of sex judgments will allow us to explore this possibility.

Enlow (1982) has suggested that the difference between male and female faces is related to the differences between infant and adult faces (see also Alley, 1988), which are in turn related to the differences between leptoprosopic and euryprosopic faces outlined above. Particularly, he has suggested that male and female faces (cf. infant and adult faces) differ in the region of the nose and brow because of requirements for increased air passages to support the relatively larger lung capacity of the male. Male faces, according to Enlow, have larger noses and more protruberant brow regions, and therefore have a convex facial profile, while female faces have smaller noses and brows that result in a concave or "dished" facial profile. Certainly, our results (Roberts & Bruce, 1988) were consistent with this proposal. In our current work we have explored this further by examining the effects of changes to the nose on the judgments made of the age and sex of our three-dimensional head models (Bruce, Doyle, Burton, & Dench, in preparation).

Using the same descriptions of heads as described above, it is possible to make local as well as global transformations. Using spline techniques, one may define the region over which a transformation has its effect. To construct transformed heads for the following experiments, the nose shape was modified by altering the profile of the greatest vertical slice through the nose, viewed in the median or sagittal plane, and then blending this change in with the rest of the face in each of the affected horizontal slices. The section of the nose bridge to be altered was defined by those points where the nose profile starts a curve to blend in with the forehead and the tip. In each case the precise point was where the slope of the curve passed through zero. The new profile was defined by a fourth-order Beta-spline curve with five control points. The start and finish of the section had one control point each, and the distance between them was divided into three equal parts with a control point at each boundary. The only point to move was the central one, and the other two lying in the tangent to the curve ensured that the final shape blended exactly into the existing curve at either end. The blending in of horizontal slices was defined by a third-order beta spline curve with four control points placed so that the ensuing curve blended in to the face surface on either side of the nose and to the tangent at the highest part of the nose bridge.

In our first experiment, we showed subjects different versions of two basic head models, that of a woman aged 35 when the laser scan was taken, and a man aged 29. These heads were shown in profile and 3/4 views, with noses pulled in (retrousse), normal, or pushed out (hooked) (see Figure 10.3). Subjects were asked to judge the age (in years), and rate the apparent masculinity/femininity, and attractiveness of the resulting heads. Our results confirmed Enlow's predictions. There was a significant effect of the nose shape on both the apparent age and the apparent femininity of the face, with hooked noses giving rise to older and more masculine heads than normal and retrousse noses. The hooked noses were also seen as less attractive. Interestingly, subjects were quite accurate in their judgments of the apparent ages of the heads, seeing the woman as 34 years on average and the man as 26 years on average, suggesting that the "baldness" of the models is not seriously distorting the perceived age levels of these heads.

In a second experiment, we set out to vary nose shape and cardioidal strain level independently. Recall that we had found cardioidal strain to yield an unreliable indication of relative age in naive subjects, yet our naive subjects in the

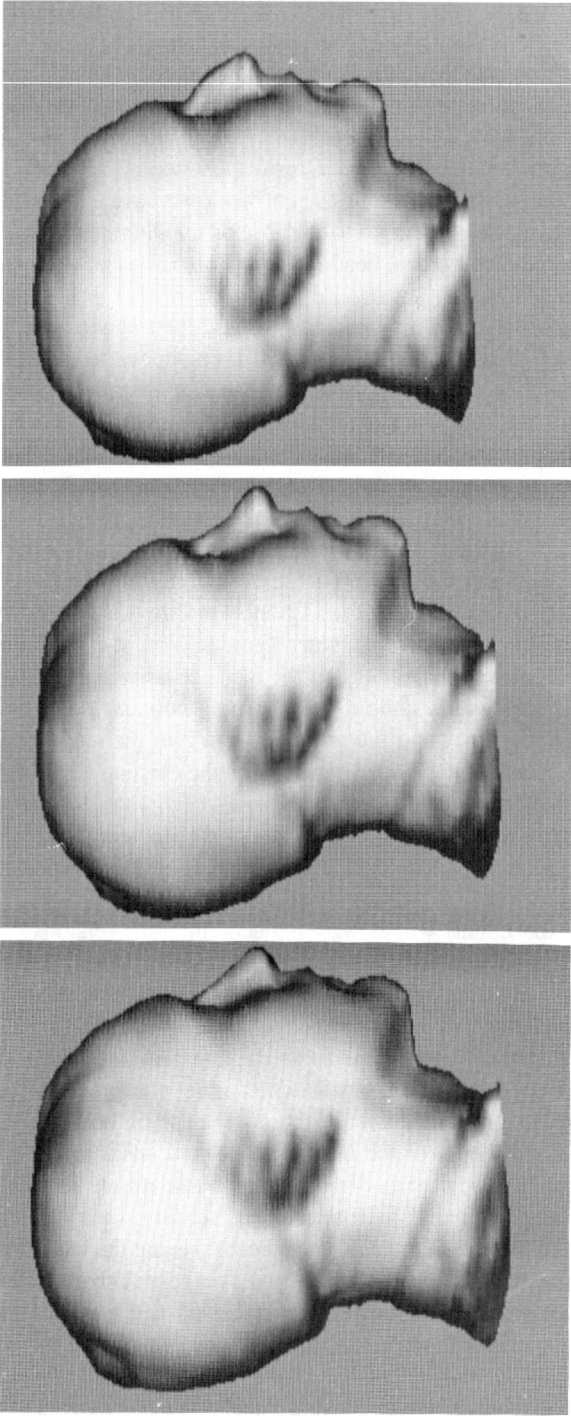

Figure 10.3. Local changes to the face surface. The changes to the nose have been blended into a small region.

experiment just described gave appropriate absolute age judgements to the heads with varying noses. Using the basic male and female head models, we presented the retroussee and hooked nose versions at three different age levels (k = 0, k = −0.14, and k = −0.28), in profile and 3/4 views, to subjects who were asked again to judge the age, sex, and attractiveness of the resulting faces (see Figure 10.4). There was a significant and reliable effect of nose shape on judgments of age and sex: retrousse noses led to younger ages and reduced masculinity compared with hooked noses. However, the effect of cardioidal strain was again inconsistent. Heads which should look "younger" were actually judged as significantly older on average, though cardioidal strain had reliable effects in the *right direction* on apparent sex of the face. A number of subjects saw faces to the right Figure 10.4 as belonging to old ladies, and those to the left as belonging to young men. Paradoxically, then, we have found greater support for the impact of cardioidal strain level on the judgment of sex which is seen as related to that of age than we have on judgments of age itself.

Conclusions on Judging Age and Sex

The experiments we have summarized above lead to a number of conclusions. First, it appears to make sense to suggest, like Mark and Todd (1983), that the visual system extracts a global shape invariant, cardioidal strain, *in three dimensions*, although we have also shown that the interpretation of levels of this transformation is sensitive to "set." Second, we have shown that changing the shape of the profile of the nose has a reliable effect on two individuating charactersitics of a face—its apparent age and sex. Note that the nose is a feature which is usually reported as not salient, and indeed changes in it may go undetected in typical experiments on feature salience from full face images (e.g., Haig, 1986a). Alley and Hildebrandt (1988) note the paucity of research into the social impact of variations in nose shape, given the frequency of surgical alterations to this region of the face. Our research would support the view that altering the shape of the nasal profile could certainly have an impact on the social impression made by the face.

While these studies have thus yielded some useful insights, they do not in themselves address the issue of whether surface-based descriptions are important to specify individual *identity*, since we have already noted that the assessment of age and sex may proceed via an independent processing route. Certainly, representations such as those shown in Figures 10.2 and 10.3 are not sufficient for accurate identification of the person shown. In a preliminary experiment, 50 subjects were shown the 3/4 view of the head model of the 35-year-old woman alongside full-face photographs showing this woman with three other women of similar age; 29 of these subjects (58%) picked the correct photograph of the woman whose head was shown, which was a rate substantially above chance

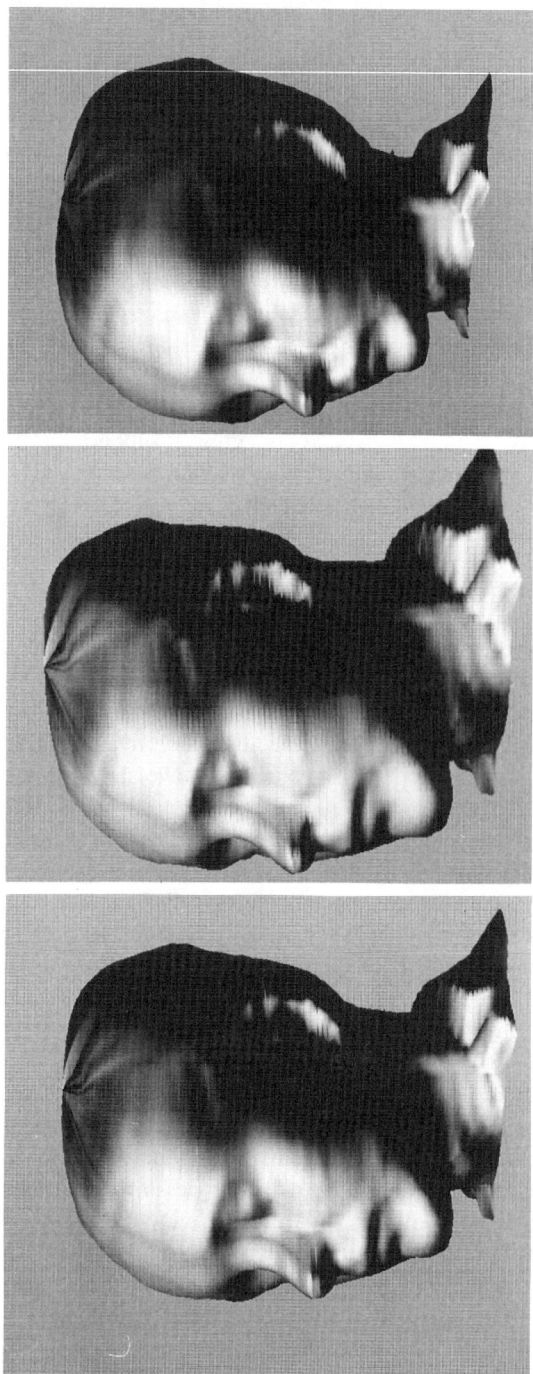

Figure 10.4. Combination of local and global transformations. In these pictures, the images have been transformed in the "younger" direction of cardioidal strain, and have retrousse noses.

(25%) but far from impressive. We are currently conducting extensions of this work using other target heads, and comparing their presentation in the "bald" state shown here, with the addition of a generic hairstyle.

This assessment of the identifiability of unknown faces from a 3-D featureless model is important in a forensic context, where the head of an unknown deceased person may be constructed from the skull. A recent example was the construction of the head of an unidentified victim of the King's Cross fire on the London underground. We do not know how well such a model can convey identity, whether showing different viewpoints would be beneficial (a single view of the King's Cross victim was shown in the press), or whether the addition of the "wrong" hair will facilitate or depress its recognizability. Our experiments will address these issues.

At a theoretical level, does the rather miserable identification of these 3-D models invalidate the surface-based approach? We think not, since we are not arguing that faces are represented *only* as surface descriptions. We know, for example, that hairstyle is one of the most important dimensions in the recognition and discrimination of unfamiliar faces (Shepherd et al., 1981), and the very baldness of our models is disguising one vital clue to their identity. Of the internal features, our head models have their eyes closed, concealing the most important internal clue to identity. It is important to acknowledge that an internal representation of a face will be of *features* in a surface, not just of *surface* alone. While there is now very good evidence that the configuration of a face is important in addition to the individual features (e.g., Sergent, 1984; Young, Hellawell, & Hay, 1987), there is no denying that individual features are important. Our aim in the study of the surface-based approach to face recognition is to help specify this mysterious "configural" component, and to examine how the visual system may extract an understanding of the embeddedness of features in a rubbery surface rather than in the picture plane.

IMPLICATIONS OF CONNECTIONISM

Until now it has been implicit in this chapter that the human visual system recognizes faces through constructing some structural description of its input, and comparing this with some store of abstract structural descriptions characterizing different individual faces (cf. Hay & Young, 1982; Bruce & Young, 1986). Here we have suggested that we may better understand the nature of such structural descriptions by understanding and incorporating the three-dimensionality of faces into our theories.

However, with the rapid development of "connectionist" approaches to the problem of human visual recognition (e.g., McClelland & Rumelhart, 1985; Kohonen, Oja, & Lehtiö, 1981; Feldman, 1985; Rolls, this volume), some might argue that there is no need to specify the nature of the representations abstracted and constructed from faces. Somehow the human visual system extracts statistical regularities from the patterns with which it is bombarded, and that is all one needs to understand about the basis for recognition.

We have a number of different points to raise in answer to this. First, the most recent and successful demonstrations of learning within connectionist systems suggest that the visual system learns principled characteristics of its input. For example, "hidden units" in multilayered pattern classification systems appear to end up looking remarkably like "feature detectors" (Hinton, 1989). In order fully to understand the workings of a PDP model, we must still understand what the model is doing, and this will require a formal characterization of the input. Our feeling is that the connectionist approach will complement, not supplant, current, more traditional computational approaches (cf. Mayhew & Frisby, 1984; Fodor & Pylyshyn, 1988; Phillips & Smith, 1989).

Secondly, whatever the details of what the human visual system does, the witness uses an "abstracted" description of a face when describing a person's appearance. A witness will describe a person's "eyes," "nose," and "mouth" even if the visual system does not label such parts explicitly. The categories that we use to describe faces may derive from different functions subserved by the different parts (eyes look, noses smell, mouths speak) rather than from any priority in visual processing, but a mapping must exist at some level between the primitives used by the visual system and those used at a more cognitive level. It is not clear how simple pattern-associators would accommodate such mappings.

Finally, even if we were to accept, for sake of argument, that no level of formal description is achieved by the visual system, and that our cognitive categories simply emerge from the juxtaposition of different associative contingencies (cf. Phillips & Smith, 1989), understanding the nature of the input and the goals of perception may provide important directions to understanding how such categories may be learned. Here we would like to propose that the mobility of the facial surface might actually *enhance* the encoding of the second-order configural variations that help specify identity.

Until fairly recently, psychologists (e.g., Hay & Young, 1982; Bruce & Young, 1986) have been content to demonstrate how the perception of expressions and lipreading appears to proceed *independently* of the recognition of individual identity. Support for the notion of independent modules devoted to different aspects of face perception comes from experiments with normal subjects (e.g., Bruce, 1986; Young, McWeeny, Hay, & Ellis, 1986), neuropsychological dissociations (e.g., Bruyer et al., 1983; Kurucz & Feldmar, 1979) and neurophysiological evidence that different cells in the monkey brain appear sensitive to identities and expressions (e.g., Rolls, 1984, 1988). The observed independence of these processes one from another is logical, given their differing task demands. To recognize an expression involves recognizing, say, a smile, despite all the different facial identities bearing it; while to recognize an identity involves recognising a particular set of features in a particular configural relationship with each other, despite all the differences in momentary configuration from one expression to the next.

However, it is interesting to note that, to the extent that a particular facial identity is learned in any simple pattern-associating manner, the "noise" present in the set of learning patterns as expressions and viewpoints vary could actually enhance the encoding of the "prototype" of this set. It is a property of some PDP pattern associators learning to discriminate nonorthogonal patterns that learning the prototype pattern of each pattern class (e.g., each different face identity) is actually improved when the exemplars of each category vary, compared with when they remain constant. We have recently discovered that people are remarkably good at learning the prototype configuration of a face from a very few, slightly different variations of the prototype.

In our experiments (Bruce, Doyle, Dench, & Burton, 1991), subjects were shown four slightly different versions of each of 10 or more different individual faces (e.g., see one set of faces in Figure 10.5) and were asked to estimate the age and rate the apparent masculinity of each studied instance (this part of the study was prompted by our research on the perception of age and sex from 3-D images of faces, as discussed earlier in the chapter). The four different exemplars of each facial identity were created by varying the vertical placement of the features. For example, in some experiments the features were shifted down as a group by 3, 6, 12, or 15 pixels. The missing prototype of this series would be a face whose pixels were shifted down by 9 pixels. Having rated these faces, subjects were given an unexpected forced choice memory task, where they were asked to choose the "old" faces from a pair showing the shifted 0 version (not previously seen) and the shifted 9 version (not previously seen either). Subjects chose the shift 9 version on a substantial majority of trials (averaging 82% in one experiment). That such "implicit" learning of the prototype is not just a *general* property of memory was demonstrated in further experiments showing that the prototype was responded to less strongly if pictures of houses were studied whose internal features (windows) were similarly shifted to create the different exemplars, or if faces (or houses) were studied and tested in *inverted* images. It seems that this prototype extraction is sensitive to the "meaningfulness" of the configural changes and/or the materials being studied.

Subtle differences in the momentary configuration of a face are important to us, since it is these which determine the expression, or speech sound, that the face is producing. Other such postural changes may be important in telling us whether the person is looking at us or not—or whether they are nodding in agreement. It is therefore important that we notice such configural variations. Given that such configural changes must therefore be *encoded* at some level, it may be that the encoding of these variations about the basic configuration of the person's face *enables* the details of his or her, particular facial "prototype" to be assimilated.

a

b

d   c   e

Figure 10.5. Slight variations (a,b,d,e) of the facial prototype (c). The patterns were created using Mac-a-Mug Pro™. (see Shepherd & Ellis, this volume.) Mac-a-Mug Pro™ is a trademark licensed to Shaherazam.

CONCLUDING REMARKS

Understanding the nature and movements of facial surfaces has clearly been essential for the development of some of the animation and reconstructive techniques described in the second part of this book. Here we have suggested that understanding the nature of faces, the functions subserved by face perception, and their interrelationships may also have an important influence on theories of the representation of the human face by the human brain.

Such an understanding may also have forensic implications, and we have already mentioned the possibility of endowing an electronic composite system with some "intelligence." Additionally, changing the shape of such features as the nose or jaw may be achieved more naturally using a 3-D representation. The techniques developed by such groups as Duffy's could allow a reconstruction of a face to be animated, shown in different views, or shown speaking a particular sentence. In this part of the book, Shepherd and Ellis described the FRAME system—an excellent example of how psychology and technology can combine to produce an aid to witness identification. It will indeed be interesting to examine the potential of the 3-D technology explored here to help develop further aids to such processes.

REFERENCES

Alley, T.R. (Ed.). (1988). *Social and applied aspects of perceiving faces.* Hillsdale, NJ: Erlbaum.

Alley, T.R., & Hildebrandt, K.A. (1988). Determinants and consequences of facial aesthetics. In T.R. Alley (Ed.), *Social and applied aspects of perceiving faces.* Hillsdale, NJ: Erlbaum.

Bassili, J.N. (1978). Facial motion in the perception of faces and in emotional expression. *Journal of Experimental Psychology: Human Perception and Performance, 4*, 373–379.

Bassili, J.N. (1979). Emotion recognition: The role of facial movement and the relative importance of upper and lower areas of the face. *Journal of Personality and Social Psychology, 37*, 2049–2058.

Brennan, S.E. (1985). Caricature generator: Dynamic exaggeration of faces by computer. *Leonardo, 18*, 170–178.

Bruce, V. (1986). Influences of familiarity on the processing of faces. *Perception, 15*, 387–397.

Bruce, V. (1988). *Recognising faces.* London: Erlbaum.

Bruce, V., Burton, M., Doyle, T., & Dench, N. (1989). Further experiments on the perception of age in three dimensions. *Perception & Psychophysics, 46*, 528–536.

Bruce, V., Doyle, T., Burton, M., & Dench, N. (in preparation). *Effects of global and local facial transformations on judgments of age and sex.* Manuscript in preparation.

Bruce, V., Doyle, T., Dench, N., & Burton, M. (1991). Remembering facial configurations. *Cognition,* in press.

Bruce, V., Ellis, H.D., Gibling, F., & Young, A. (1987). Parallel processing of the sex and familiarity of faces. *Canadian Journal of Psychology, 41,* 510–520.

Bruce, V., & Valentine, T. (1988). When a nod's as good as a wink: The role of dynamic information in facial recognition. In P.E. Morris, M.M. Gruneberg, & R.N. Sykes (Eds.), *Practical aspects of memory: Current research and issues* (Vol 1, 169–174).

Bruce, V., Valentine, T., & Baddeley, A. (1987). The basis of the 3/4 view advantage in face recognition. *Applied Cognitive Psychology, 1,* 109–120.

Bruce, V., & Young, A. (1986). Understanding face recognition. *British Journal of Psychology, 77,* 305–327.

Bruyer, R., Laterre, C., Seron, X., Feyereisen, P., Strypstein, E., Pierrard, E., & Rectem, D. (1983). A case of prosopagnosia with some preserved covert remembrance of familiar faces. *Brain and Cognition, 2,* 257–284.

Davies, G.M. (1986). The recall and reconstruction of faces: Implications for theory and practice. In H.D. Ellis, M.A. Jeeves, F. Newcombe, & A. Young (Eds.), *Aspects of face processing.* Dordrecht, Netherlands: Martinus Nijhoff.

Davies, G.M., Ellis, H.D., & Shepherd, J.W. (1978). Face recognition accuracy as a function of mode of representation. *Journal of Applied Psychology, 63,* 180–187.

Ellis, H.D. (1986). Processes underlying face recognition. In R. Bruyer (Ed.), *The neuropsychology of face perception and facial expression.* Hillsdale, NJ: Erlbaum.

Ellis, H.D., Davies, G.M., & Shepherd, J.W. (1978). A critical examination of the Photofit system for recalling faces. *Ergonomics, 21,* 297–307.

Enlow, D.H. (1982). *Handbook of facial growth.* Philadelphia, PA: W.B. Saunders.

Feldman, J.A. (1985). Four frames suffice: A provisional model of vision and space. *The Behavioural and Brain Sciences, 8,* 265–289.

Fodor, J.A., & Pylyshyn, Z.W. (1988). Connectionism and cognitive architecture: A critical analysis. *Cognition, 28,* 3–71.

Gibson, J.J. (1979). *The ecological approach to visual perception.* Boston, MA: Houghton Mifflin.

Haig, N.D. (1986a). Investigating face recognition with an image-processing computer. In H.D. Ellis, M.A. Jeeves, F. Newcombe & A. Young (Eds.), *Aspects of face processing.* Dordrecht, Netherlands: Martinus Nijhoff.

Haig, N.D. (1986b). Exploring recognition with interchanged facial features. *Perception, 15,* 235–247.

Hay, D.C., & Young, A.W. (1982). The human face. In A.W. Ellis (Ed), *Normality and pathology in cognitive functions.* London: Academic Press.

Hinton, G. (1989). connectionist learning procedures. *Artificial Intelligence, 40,* 185–234.

Kohonen, T., Oja, E., & Lehtio, P. (1981). Storage and processing of information in distributed associative memory systems. In G. Hinton & J.A. Anderson (Eds.), *Parallel models of assciative memory.* Hillsdale, NJ: Erlbaum.

Kurucz, J., & Feldmar, G. (1979). Prosopo-affective agnosia as a symptom of cerebral organic disease. *Journal of the American Geriatrics Society, 27,* 225–230.

Laughery, K.R., & Fowler, R.H. (1980). Sketch artist and Identikit procedures for recalling faces. *Journal of Applied Psychology, 65,* 307–316.

Mark, L.S., Pittenger, J.B., Hines, H., Carello, C., Shaw, R.E., & Todd, J.T. (1980). Wrinkling and head shape as coordinated sources of age-level information. *Perception & Psychophysics, 27*, 117–124.

Mark, L.S., & Todd, J.T. (1983). The perception of growth in three dimensions. *Perception & Psychophysics, 33*, 193–196.

Mark, L.S., & Todd, J.T. (1985). Describing perceptual information about human growth in terms of geometric invariants. *Perception & Psychophysics, 37*, 249–256.

Mark, L.S., Todd, J.T., & Shaw, R.E. (1981). Perception of growth: A geometric analysis of how different styles of change are distinguished. *Journal of Experimental Psychology: Human Perception and Performance, 7*, 855–868.

Marr, D. (1976). Early processing of visual information. *Philosophical Transactions of the Royal Society of London, B275*, 483–524.

Mayhew, J., & Frisby, J. (1984). Computer vision. In T. O'Shea & M. Eisenstadt (Eds.), *Artificial intelligence*. New York: Harper & Row.

McClelland, J.L., & Rumelhart, D.E. (1985). Distributed memory and the representation of general and specific information. *Journal of Experimental Psychology: General, 114*, 159–188.

Pearson, D.E., & Robinson, J.A. (1985). Visual communication at very low data rates. *Proceedings of the IEEE, 73*, 795–812.

Phillips, W.A., & Smith, L.S. (1989). Conventional and connectionist approaches to face processing by computer. In A.W. Young & H.D. Ellis (Eds.), *Handbook of research on face processing*. Amsterdam, Netherlands: North-Holland.

Pittenger, J.B., & Shaw, R.E. (1975a). Aging faces as viscal elastic events: Implications for a theory of nonrigid shape perception. *Journal of Experimental Psychology: Human Perception and Performance, 1*, 374–382.

Pittenger, J.B., & Shaw, R.E. (1975b). The perception of relative and absolute age in facial photographs. *Perception & Psychophysics, 18*, 137–143.

Rhodes, G., Brennan, S.E., & Carey, S. (1987). Identification and ratings of caricatures: Implications for mental representations of faces. *Cognitive Psychology, 19*, 473–497.

Roberts, T., & Bruce, V. (1988). Feature saliency in judging the sex and familiarity of faces. *Perception, 17*, 475–481.

Rolls, E.T. (1984). Neurons in the cortex of the temporal lobe and in the amygdala of the monkey with responses selective for faces. *Human Neurobiology, 3*, 209–222.

Rolls, E.T. (1988). Visual information processing in the primate temporal lobe. In M. Imbert (Ed.), *Models of visual perception: From natural to artificial*. Oxford, England: Oxford University Press.

Sergent, J. (1984). An investigation into component and configural processes underlying face recognition. *British Journal of Psychology, 75*, 221–242.

Shepherd, J.W., Davies, G.M., & Ellis, H.D. (1981). Studies of cue saliency. In G. Davies, H. Ellis, & J. Shepherd (Eds.), *Perceiving and remembering faces*. London: Academic Press.

Ullman, S. (1979). *The interpretation of visual motion*. Cambridge, MA: M.I.T. Press.

Valentine, T., & Bruce, V. (1986). The effects of distinctiveness in recognising and classifying faces. *Perception, 15*, 525–536.

Watt, R.J. (1988) *Visual processing: Computational, psychophysical and cognitive research*. London: Erlbaum.

Young, A.W., Hellawell, D., & Hay, D.C. (1987). Configurational information in face perception. *Perception, 16,* 747–760.

Young, A.W., McWeeny, K.H., Hay, D.C., & Ellis, A.W. (1986). Matching familiar and unfamiliar faces on identity and expression. *Psychological Research, 48,* 63–68.

Author Index

Subject Index